D0863257

Algrove Publishing Limited
1090 Morrison Drive
Ottawa, Ontario
Canada K2H 1C2

Canadian Cataloguing in Publication Data

Fair, Albert
Shortcuts in carpentry : a collection of useful problems giving new and improved methods of laying out and erecting carpenters' work

2nd ed.
(Classic reprint series)
Includes index.
Includes original t.p.: Shortcuts in carpentry : a collection of useful problems giving new and improved methods of laying out and erecting carpenters' work / by Albert Fair.
New York : Industrial Book Co., 1909.
ISBN 0-921335-45-8

1. Carpentry. 2. Architectural woodwork. 3. Laying-out (Woodwork). I. Title. II. Series: Classic reprint series (Ottawa, Ont.).

TH5604.F33 1998 694 C98-900999-8

Printed in Canada
#21299

C.W. Frost

Publisher's Note

Problems in carpentry are much the same today as they were 90 years ago when this book was written. The only differences today's carpenter will notice will be in the items being made; the principles involved are timeless.

This edition is exactly the same as the original 1909 edition, except for the placement of the fold-out drawing and the repetition of the related terms. The original layout was not nearly as convenient to view.

Leonard G. Lee, Publisher
Ottawa
September, 1998

SHORT CUTS IN CARPENTRY

A COLLECTION OF USEFUL PROBLEMS

GIVING

NEW AND IMPROVED METHODS OF LAYING OUT AND ERECTING CARPENTERS' WORK

BY

ALBERT FAIR

Author of "Steel Square as a Computing Machine," "Practical House Framing," etc., etc.

WITH NUMEROUS ILLUSTRATIONS

NEW YORK

INDUSTRIAL BOOK COMPANY

1909

SHORT CUTS IN CARPENTRY

CONTENTS

PREFACE

QUITE a number of carpenters pride themselves for knowing a number of short cuts, and quite often they brag about it a good deal.

When the ones that brag the loudest are given a job a little out of the ordinary, they fall flat—they simply don't know how to go about it.

On the other hand there is the carpenter who understands the principles of his trade, and he rarely is at a loss of knowing how to lay out a job. On certain kinds of work he is very slow compared with the man who knows the short cuts for that kind of work, but his slowness is more than made up by his steady application. You don't see this sort of a "chip" look at a simple drawing half a day, wondering how he should commence the work, as some of those "lightning speeders" do.

In the preparation of this book it has been the aim of the compiler to give the principle of the various subjects, so that the reason for the "short cut" will be understood. By this plan it is the author's wish that the young carpenter will first learn how to lay out work correctly, and second, how to do it quickly.

Considerable of the matter given in this book has been contributed to the columns of *The Practical Carpenter* by various writers.

In arranging this matter for publication in book form, considerable new and original matter has been added.

Proper credit is given to all whose practical hints have been used, and where credit is omitted, the article is original with the editor.

The publishers and editor believe that this book will prove o real value to carpenters, by enabling them to save time in laying out and erecting work.

To the young man especially it is hoped it will be useful, as it gives the results of years of experience, and to make the path easier for the earnest young man is always the desire of

ALBERT FAIR.

NEW YORK, Nov. 1, 1908.

SHORT CUTS IN CARPENTRY

THE information given in this book is intended for carpenters in general, whether they be carpenters, joiners, or mill woodworkers.

The subject of whether a workman is a carpenter or joiner, or both, is we l explained by John Black as follows:

The phrase "a carpenter and joiner" is in sufficiently common use, and to the ordinary non-technical mind is not infrequently supposed to refer to two lines of work almost, if not entirely, identical. But this is entirely erroneous. True, both the carpenter and the joiner are woodworkers, and each uses the saw, the plane, the chisel, etc. But there the resemblance stops. It is in the size of the "stuff" which they deal with that the distinction comes in, and this is better shown in the analogous French terms. In that language a *charpentier* is a man who uses timber of fair scantling for the construction of roof trusses, partitions, or timber houses, while the *menusier*, as his title imparts, uses small—*menu*—stuff for doors, stairs, windows, and other fittings.

In the early period of the history of the building art, the labor involved in the construction of edifices formed solely or mainly of wood was undertaken by one class of workmen only. The heavy timber framing, the massive roof, the thick planks which formed the flooring, and the rude iron-studded door required no delicate manipulation, no elaborate tools; but by the rough aid of saw, axe, and shave, were brought to whatever comeliness they possessed. In these old edifices the carpenter was the only artisan needed. It was not until ideas of comfort and luxury had made some progress that the inside fittings of the building—staircase, interior doors, wood-framed windows, and other refinements—became essential, and in the production of these lighter forms of woodwork, which necessitated the use of most varied and accurate tools, the joiner found his place as distinct from the carpenter. The latter workman constructed the heavy and bulky portions of the building, while to the joiner was left the fitting up of the lighter and ornamental parts.

One of the greatest aids to a young man new to the trade is a guide to the various technical names of the interior of a house. To fill this want the editor prepared a drawing of an interior, which practically explains itself. (See Fold-out at back)

To get this drawing within a reasonable size, so that it can be presented in this book, it was necessary to group the various designs together, so as to show the different styles of trim, etc. Some of the details are exaggerated in size compared with the scale of the drawing, so as to show them plainly. The names of the various parts are:

1. Window-sill.
2. Sub-sill.
3. Furring.
4. Quarter rounds.
5. Sill-cap.
6. Header.
7. Pocket or opening to weight box.

8. Blind-hanging stile or exterior casing.
9. Exterior sash stop.
10. Clap-boarding, shingles, or outside covering.
11. Sheathing or roof boarding.
11^1. Sheathing paper.
12. Stud.
13. Laths.
14. Plastering.
15. Architrave, interior casing, or window trim.
16. Stop bead.
17. Parting bead or strip.
18. Pulley stile.
19. Sash weights.
20. Window latch or sash lock.
21. Pulley
22. Sash cord.
23. Meeting rail of outside sash.
24. Meeting rail of inside sash.
24^1. Bottom rail of sash.
24^2. Top rail of sash.
25. Stop bead.
26. Window head.
27. Exterior casing.
28. Sash stile.
29. Astragal or sash bar.
30. Sash lift.
31. Window pane or glass.
32. Panel back or breast.
33. Base block .

34. Window trim, casing, or architrave.
35. Corner block.
36. Window trim.
36[1]. Cap trim for window.
37. Picture molding.
38. Dado.
39. Wall.
40. Border.
41. Stile.
42. Hanging stile.
43. Top rail.
44. Middle rail.
45. Lock rail.
46. Bottom rail.
47. Muntin.
48. Panels.
48[1]. Upper panel.
48[2]. Middle panel.
48[3]. Lower panel.
49. Knob.
50. Keyhole.
51. Base blocks.
52. Door trim.
53. Head block.
54. Door trim.
55. Cap trim for door.
56. Carving.
57. Hinges.
58. Ornamental casing.

59. Door saddle or threshold
60. Header.
61. Door stop or jamb mold.
62. Door jamb.
63. Furring.
64. Door jamb.
65. Plaster cornice.
66. Frieze.
67. Plate rail.
68. Wainscoting.
69. Wall covered with paper.
70. Skirting or base board.
71. Chair rail.

THE USE OF GEOMETRY

Many mechanics dislike geometry, thinking it is no use to study it, as they never have to erect perpendiculars, bisect lines and angles, etc. In this they are mistaken, as geometry is of every-day use, and its principles can be applied constantly.

The application to practical work of the problem of erecting a perpendicular to a given line at a given point in that line will show how a knowledge of geometry is valuable in every-day work.

Let AB, Fig. 1, be the given line and C the point from which the perpendicular to AB is to be raised. Set off on each side of the point C the points D and E at equal distances from the point C, then from the points D and E, and with a radius greater han the distance DC or CE, draw arcs intersecting each other at F. A line drawn through the point F to the point C will be perpendicular to the line AB. Likewise the distances DF and FE are equal.

Fig. 2 shows the application of this problem to placing a piece of timber so that it will be at right angles to another piece. Measure off from each side of the upright timber any convenient

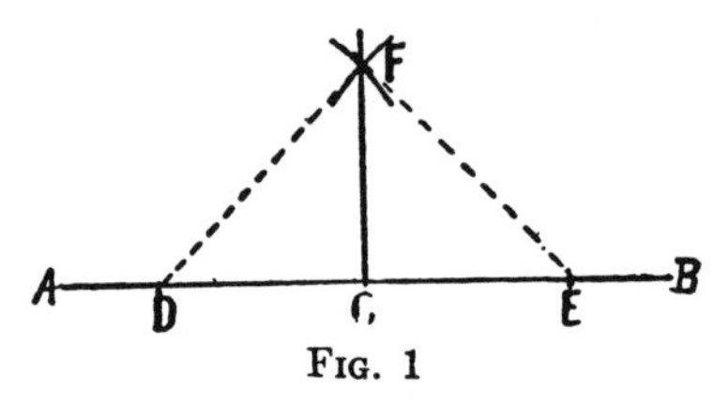

FIG. 1

distance, thus getting the points D and E. Then take two poles or slats of equal length, have these held steadily at the points D and E, and then rack the upright timber until the tops of the

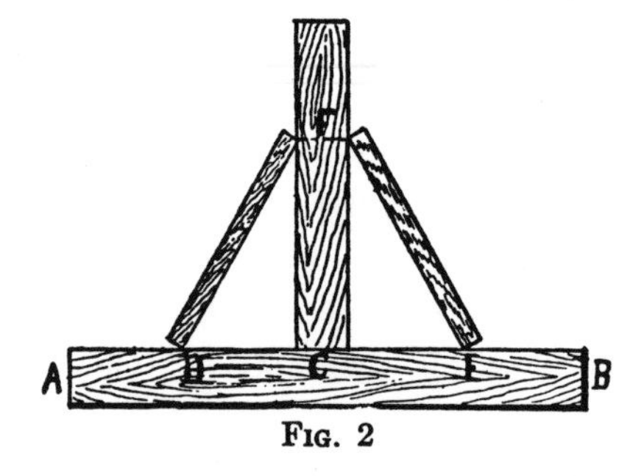

FIG. 2

poles are exactly opposite, as shown at F; when this occurs, the upright timber is exactly perpendicular to the timber AB.

Two problems that are well known are the bisecting of a straight line and the bisecting of an angle.

To bisect a line. Let AB, Fig. 3, be the given line; then

with A and B as centres, with a radius greater than one-half of AB draw arcs; these will cut each other at the points C and D; the line CD drawn through the points C and D divides the line AB into two equal parts.

To bisect an angle. Let ACB, Fig. 4, be the given angle; from C as a centre, with any radius describe an arc AB cutting the two sides of the angle at A and B; from these two points as centres, with any radius greater than one-half of AB draw two arcs cutting each other as at D; the line CD drawn through

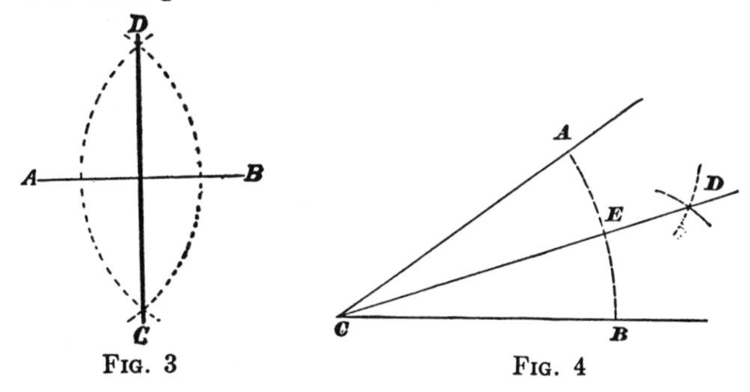

FIG. 3

FIG. 4

the vertex of the angle C and the point D divides the angle ACB into two equal parts.

MITRING

Any one understanding the foregoing geometrical problems need never be at a loss of knowing how to mitre any straight moldings together, no matter at what angles they may be.

Thus for a square mitre the bevel is quickly found; it is the line CD, Fig. 5; here the right angle ACB has been simply bisected, giving at once the required angle which can be transferred to the work by means of a bevel.

The same is true of any other mitre. Thus Fig. 6 shows an

octagon mitre. The mitre for the common regular polygons is well known, but the above problems may be used for finding the mitre of any angle.

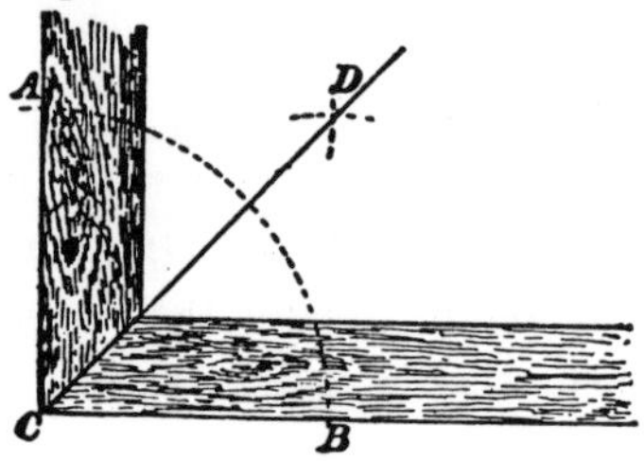

Fig. 5

Where a mitre has to be cut a number of times, it is best to have a mitre box, as its use saves time and insures accuracy.

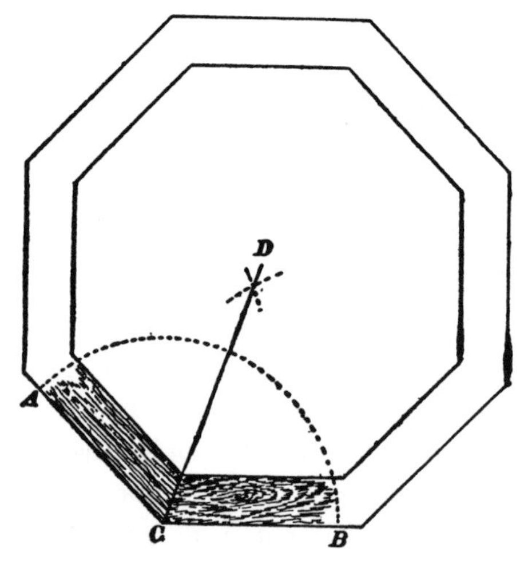

Fig. 6

On the following pages simple directions for the construction of a serviceable mitre box are given.

HOW TO MAKE A MITRE BOX

Although the construction of this labor- and time-saving device seems at a glance to be very simple, a great deal of care must be exercised in the making of a mitre box.

But very few tools are required in making a mitre box, and even the handyman about the house generally has the requisites.

Below are given two methods of making a mitre box. The object of this is to show that a box can be made with different tools.

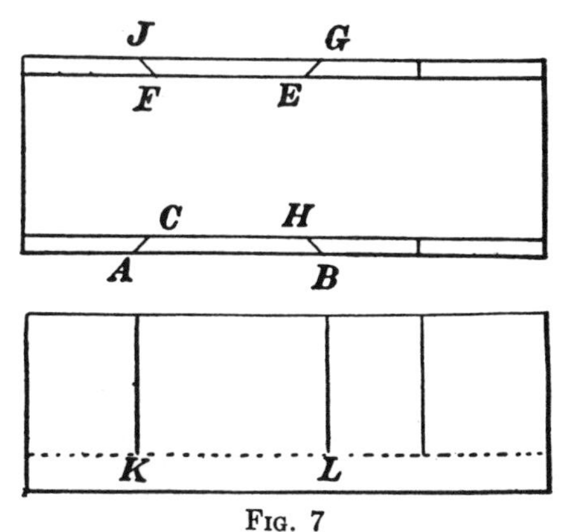

FIG. 7

Fig. 7 shows a full working drawing of a mitre box. The simplest way of making a mitre box is with the use of a mitre try-square (Fig. 8). This tool has been on the market for a number of years; room should be made for it in every carpenter's tool chest. With the use of this tool, a saw, and a hammer and nails, any handy person can make a picture frame.

For the ordinary mitre box, take three boards, the two for the sides being $\frac{1}{2}'' \times 15'' \times 5''$; the bottom piece should be

1"×15"×4". The box when put together will be 5 inches wide from side to side. Now mark off 5 inches, AB, on any part of one of the sides (Fig. 7). Then place the mitred try-

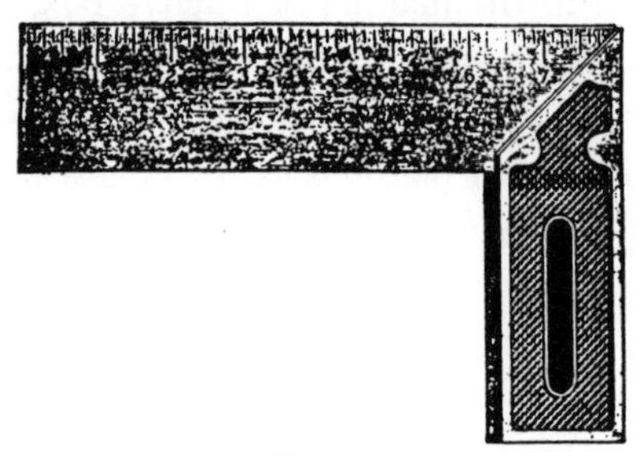

FIG. 8

square at A and mark from A to C, E to G. Also from B to H, F to J. Now place the square on A and mark side to K; also

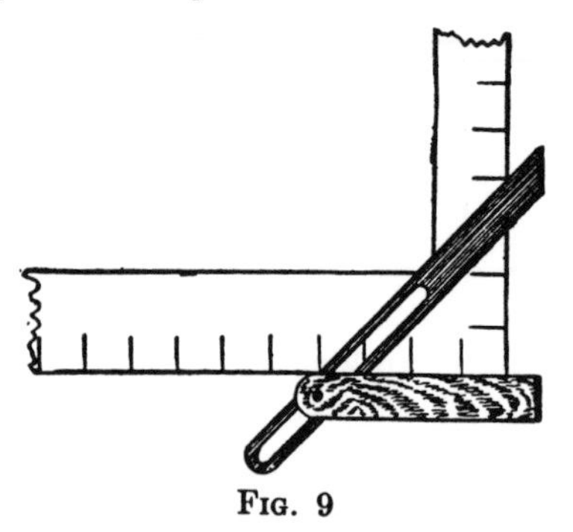

FIG. 9

from B to L. The same operation is employed on the other side from the points G and J.

The box is now ready for the sawing of the mitre cuts, with which much care must be taken.

Another method of making a mitre box is with the use of the steel square and bevel. The same material is used as in the method described above. Take a bevel and place it on the blade and tongue of a square at markings which correspond as shown in Fig. 9. (It does not necessarily have to be three and three, as shown in Fig. 9, but any numbers, as one and one, two and two, etc.) The same operation is employed with regard to marking as already described, only the bevel takes the place of the mitred try-square. The markings down the sides are done with the carpenter's steel square.

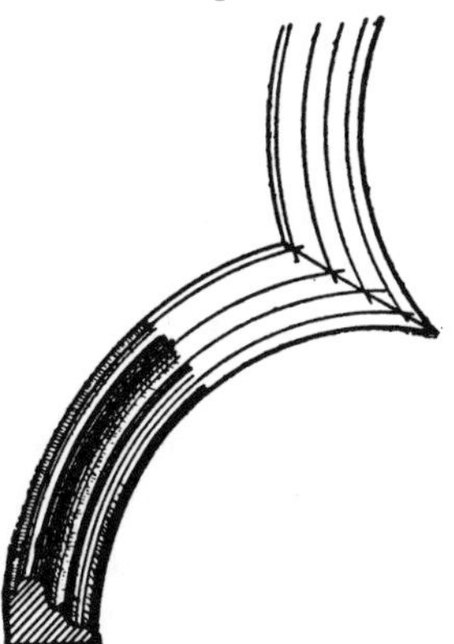

FIG. 10

MITRES FOR CURVED MOLDINGS

Frequently, especially in Gothic church work, it is necessary to mitre curved moldings. This is comparatively easy if one will remember that the mitre must be such that each part of one piece of molding must touch the corresponding part on the other piece; in other words, the mitre is at the points of intersection of the pair of parallel lines of the molding.

This will be better understood by a reference to Fig. 10, which makes the matter plain. Here it will be noticed that the mitre line is a curve instead of a straight line, as is the case with straight moldings.

MOLDINGS

When the face or edge of any work is cut or planed into long channels or projections, the sections of which form various curves, it is said to be "molded," and each separate member or group is called a molding. Moldings of various kinds are used in carpentry and joinery, their curves being guided by the taste of the designer or workman, or the material in which the work is being executed. All moldings should, however, be based on correct and standard forms, some of which are given for this purpose. Moldings are divided into Grecian, Roman, and Gothic. Grecian moldings are formed of some of the curves known as conic sections—such as the ellipse or hyperbola—and sometimes even of a straight line in the form of a chamfer. Roman moldings have their sections composed of arcs of circles, and thus they are found the easiest for elementary practice in linear drawing. Fig. 11 shows the following moldings:

The torus is the simplest of all curved moldings. It is merely a semicircle described upon a vertical diameter; it is used in the bases of columns. The torus, when very small, is called an astragal, which projects, but it is called a bead when it does not stand out beyond the surface. Several beads placed together are called reedings.

A fillet is a small, flat face placed between moldings, to divide them. A fillet is, in the bases of columns and at the top, joined to the face or to the column itself by a small quarter-

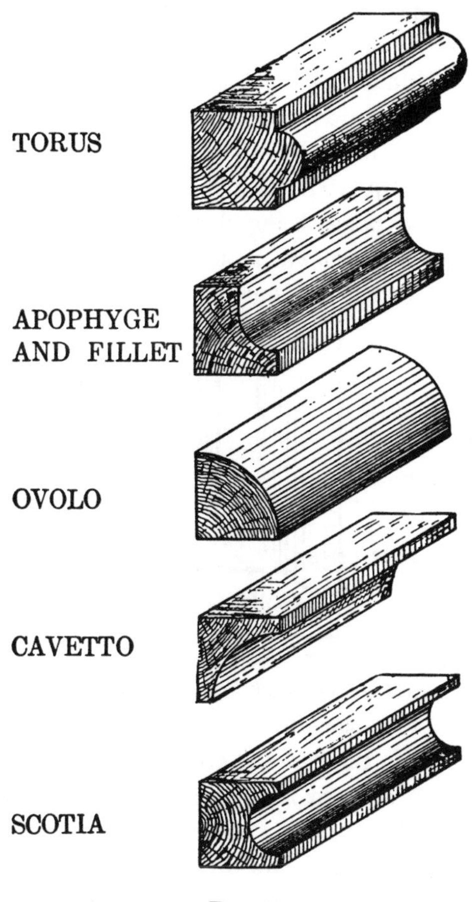

FIG. 11

round hollow, termed an apophyge. The word is originally Greek, and signifies flight. It is often called the "scape" or "spring" of a column.

The ovolo (the name of which is derived from the Latin word *ovum*, an egg) is a portion of a conic section, but in the Roman is merely a portion of a circle, generally a quadrant, in which case it is called a "quarter-round."

A cavetto, the name being derived from the Latin word *cacus*, a hollow, is a concave molding, the curvature of whose section does not exceed the quarter of a circle. Its projecting

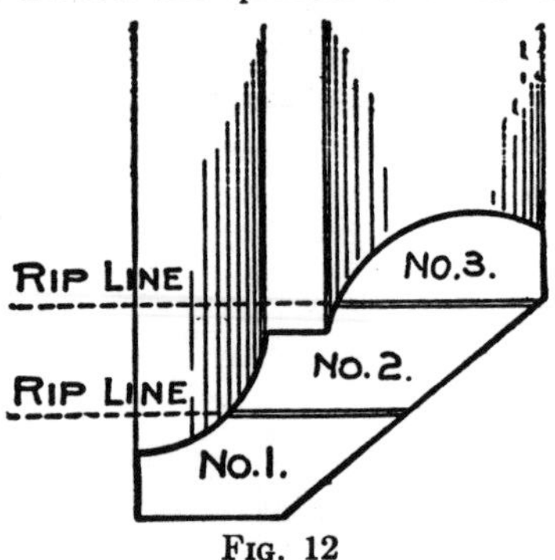

FIG. 12

depth may be equal to its height, and should never be less than two-thirds of it. The cavetto, which is the reverse of the ovolo is sometimes used in the bed and crowning moldings of cornices.

The scotia is a recessed molding of an elliptical section when properly constructed. It is, however, for general purposes, formed by the junction of two circular arcs of different radii. This molding has an effect just the opposite to that of the ovolo or torus, and is sometimes composed like the latter, of a semicircle only.

BENDING MOLDING AROUND CIRCLES

A writer in the *American Carpenter and Builder*, Walter McKay, gives a method of preparing molding to bend around circles:

Take a piece of molding and rip it into three pieces at the angle shown in Fig. 12, and then dress this strip where it has been ripped, so it will make a tight joint. Then bend one strip at a time until you have built the mold up around the circle, making a neat job, and not showing like a mold that has been kerfed.

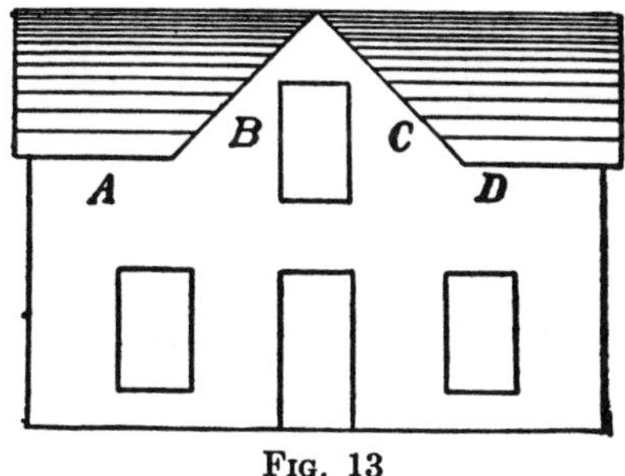

FIG. 13

RAKE MOLDINGS

One of the most difficult things for young carpenters to understand is a raking molding. In this short chapter the "whys" and "wherefores" will be explained in a fashion so that all will understand the principles.

In Fig. 13 is an outline sketch of a front elevation of a house. To make it look more ornamental, a molding is to be placed beneath the roof lines. For the sections A and D, no mention may be made, as it is simply horizontal work. The molding is to be also placed at B and C Moldings which are placed at an angle to horizontal ones are called raking moldings.

Now, if a plain board of the same width is to be used for the molding, and cut to the proper mitres and put into position, it will look like Fig. 14. A glance will show that something is wrong, although the pieces B and C are the same width as AD.

From this it is evident that the raking molding must be made from a narrower board than the horizontal molding it is to meet, so as to look like Fig. 15.

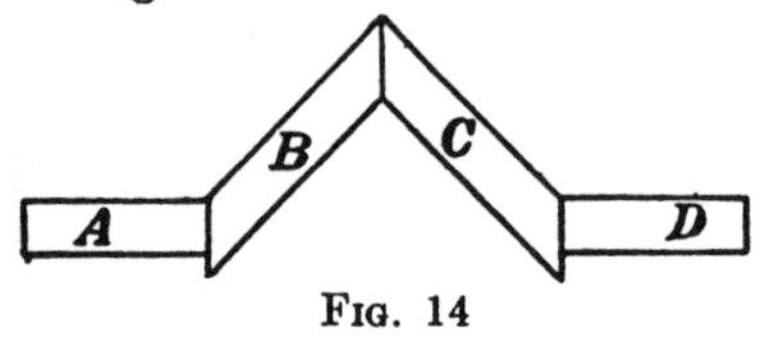

FIG. 14

The method of finding what width of board to use for the raking molding is quite simple, as the following construction will show. Draw a horizontal line, AB, Fig. 16, and on this erect a perpendicular, CD, equal to the width of the horizontal molding. (This can be done full size or to scale.) Then at

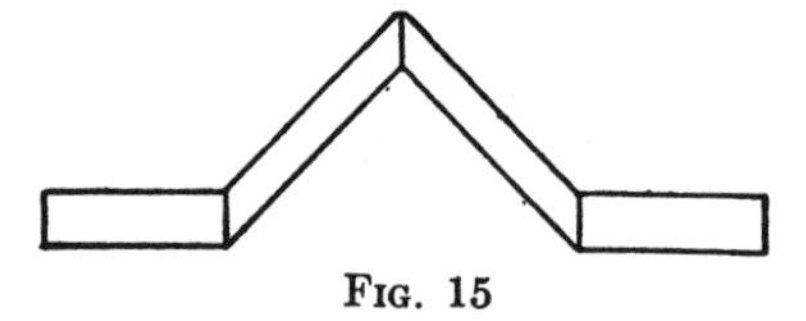

FIG. 15

the angle the rake molding is to be, draw the lines DE and CF. On the line CF, at any point, draw a perpendicular to the line DE. This distance (GH) will be the width of board.

The boards at A and D, Fig. 14, are cut plumb, but if they were cut at a mitre, as shown in Fig. 17, the width of the raking molding is found in the same manner as in Fig. 16. This is indicated by the line AB.

Moldings usually consist of more than a board, and sometimes they are highly ornamented. When the ornamental molding is used, the rake molding is found in the same way as before. Thus in Fig. 18 the left-hand side shows the section of the horizontal molding, and the right-hand side the section of the raking

Fig. 16

Fig. 17

molding. AA on the horizontal molding is equal to *aa* on the rake molding, etc. As will be easily seen, the method is to draw lines parallel to the edges of the molding from each portion of the molding to the mitre; here continue the lines parallel

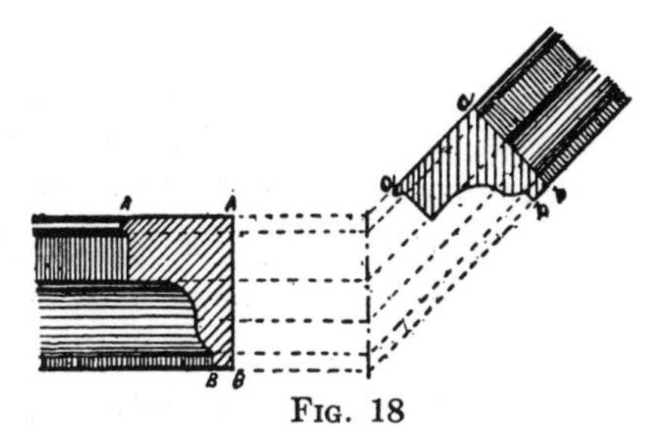

Fig. 18

to the edges of the rake molding, then draw the line perpendicular to the edges of the raking molding, and from this set off the various parts of the molding by making *aa* equal to AA, *bb* to BB, etc. In this way the section of the rake molding is found.

The above article, when published in the *Practical Carpen-*

ter, was received with considerable favor, several writers commending it and giving additional hints, Mr. O. J. Reddick writing as follows:

In Fig. 19 is shown a method of cutting rake moldings, by which the same width mold can be used both up the rake and horizontally. By Brother Albert Fair's method, a 3½-inch mold on a half-pitch roof would require a 5¼-inch mold horizontally, as shown at B, Fig. 19. By mitring as shown at A, the same width of mold may be used. This method will work on any pitch roof, but the mitres at A will not be 45-degree angles except in the half roof. The angle to cut the mitres to may be obtained by drawing a diagram and letting the horizontal

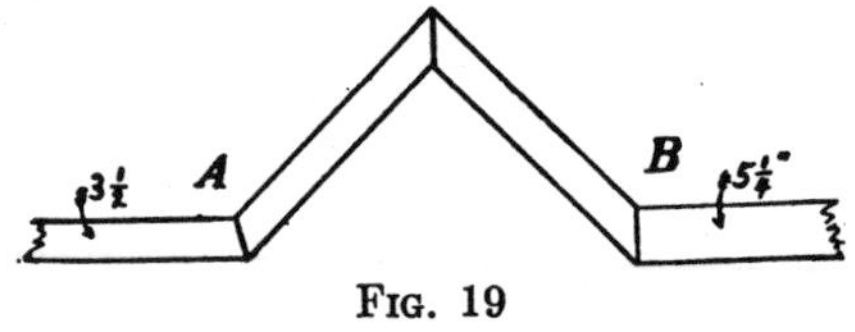

FIG. 19

and roof angle lines bisect, and then draw a line through the two points where these lines meet. This line gives the angle to which to cut both the horizontal and rake mold.

This method of obtaining the angle is quite fully explained by Mr. E. B. Bailey, who says that it works equally well on plain or ornamental work:

In laying out this work it is best to take a piece of board about 12 inches wide and 16 inches long. We will suppose the gable to be 15 feet at the base, and the roof ⅓ pitch, or 5 feet, and the horizontal and raking molding each to be 4 inches wide.

Now draw a line about 14 inches long and 1 inch from the edge of the board, as AB, Fig. 20; then draw the line CD parallel and 4 inches from AB. Draw a line perpendicular to CD, and on this set off a point, H, 5 inches from CD (5 inches is $\frac{1}{3}$ of 15 feet drawn to a scale of 1 inch to the foot). Again, beginning at D measure $7\frac{1}{2}$ inches toward C on CD, and make a dot. Now draw a line EF from E through the dot: then

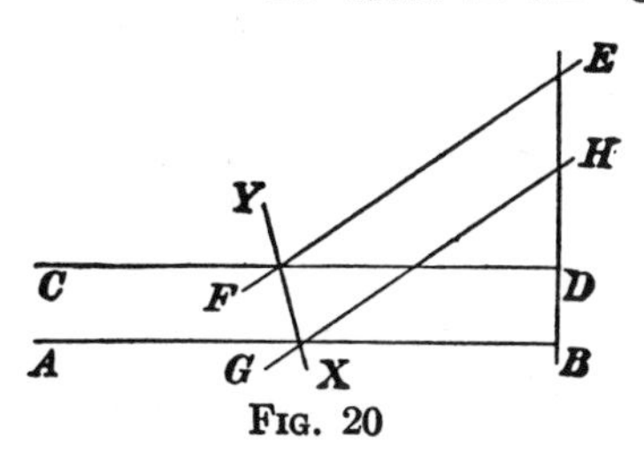

FIG. 20

draw a line GH parallel to and 4 inches from the line EF,—that is the width between the lines. Through the point of intersection of the lines CD and FE and the point of intersection of the lines AB and GH, draw the line XY, and the work is complete.

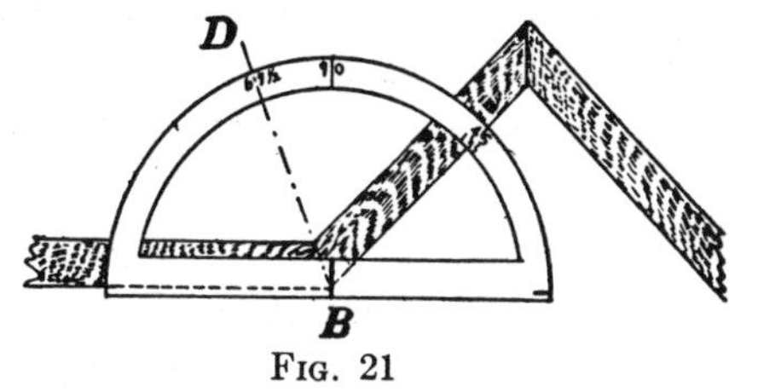

FIG. 21

Now if you place the handle of a T bevel against the edge of the board and place the blade parallel with the line XY, you can transfer the angle to your work or mitre box.

A method of using a protractor to find the mitre of a rake molding is described by E. S. Frye, which he thinks will meet the approval of all who read it, as it will work equally well on any pitch of roof:

Place the protractor upon the rake, as shown in Fig. 21, with the centre, B, at the lowest point. In a half-pitch roof place the 45° upon the run of the rake (or whatever pitch it happens to be), keeping the base horizontal. Now subtract the pitch of the roof from the 180° in the semicircle, which in this case leaves 135°. Then make the cut upon a line, BD, from centre of diameter to a point one-half the distance from 0° to 135°, which equals 67½°.

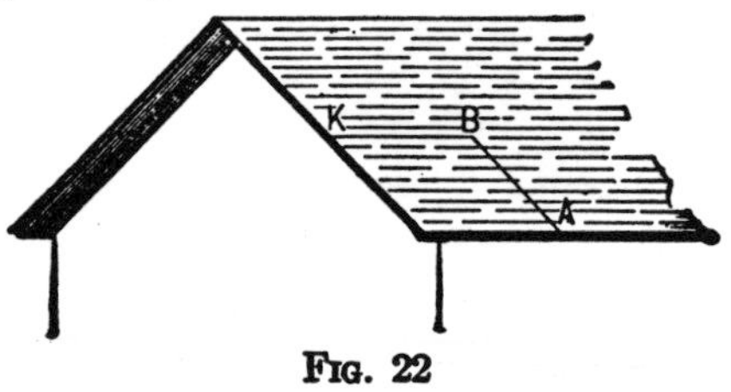

FIG. 22

Quite often the horizontal molding is on one side of the building and the rake molding on the front. In cases of this sort the molding must be mitred to make the proper joint for the cornice.

An example of this sort is shown by Fig. 22, which shows part of a gable-end roof. Two important things must be found: first, the section of the molding, and second, the mitre between the horizontal and rake molding.

For finding the section, the method shown by Fig. 16 is used. The mitre on the horizontal molding is the regular 45-degree mitre. The rake molding is to be cut at a double mitre, that is, both across and down.

A practical way is adopted for cutting this mitre by use of a mitre box.

Take an ordinary plain mitre box, IJL, Fig. 23, and draw the line AB, making the angle ABJ equal to the pitch angle of the roof, Fig. 22. Draw BD perpendicular to AB, and make it equal to the width of the box IJ; make DE parallel to AB, and extend lines from B and E square across the box to K and C; join BC and EK. ABC will be the mitre cut for two of the rake angles; HEK will be the cut for the other two angles, the angle HEN being equal to the angle ABJ. In mitring both horizontal and rake molding, that part of the molding which

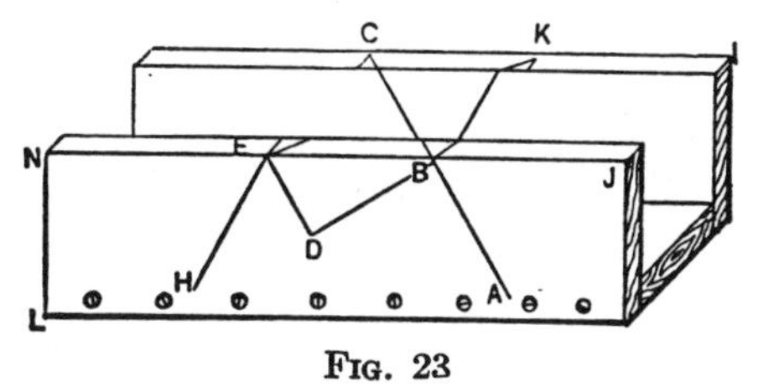

Fig. 23

is vertical when in its place on the cornice must be placed against the side of the box.

If the molding or cornice is a deep one, it will be necessary to have the horizontal molding of a certain form conforming to the pitch of the roof, as otherwise there would be a projection from the raking molding. In narrow moldings this would never be noticed, but sometimes the carpenter may be called on to do a fancy job, and it will be well to know how to find this section.

Fig. 24 shows how it is done. *klo*P*r* shows the rake molding and *acef* the form needed for the horizontal molding (the bottom and tops of which must be at the same angle as the pitch of the roof); the principles upon which the method is based being, first, that similar points on the rake and horizontal parts of a cornice are equally distant from vertical planes represented by the walls of a building, and second, that such similar points are equally distant from the plane of the roof. Represent the wall faces of a building by the line AB, and a section of the horizon-

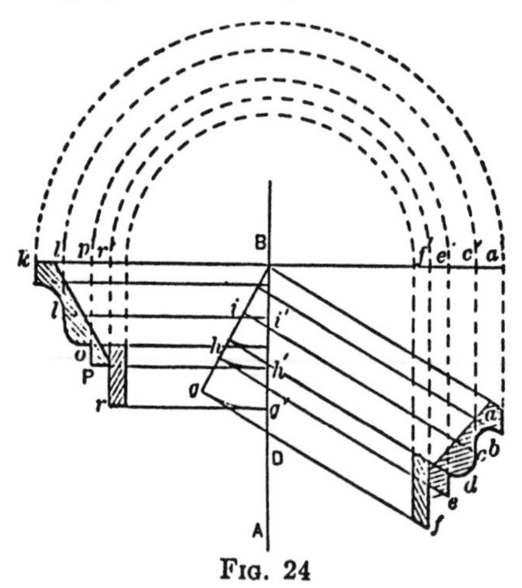

Fig. 24

tal cornice by DB, Fig. 24, which will correspond to AB and BK in Fig. 22.

First draw the vertical line AB, Fig. 24, then the line B*a* at the angle of the pitch of the roof.

Following the idea given in the above principle, draw lines *aa'*, *cc'*, ' parallel to AB and intersecting the line *ka'*, which is drawn at right angles to AB through the point B; then, with B as a centre, describe the arcs *a'k*, *c'l'*, *f'r'*, etc., intersecting the same line *ka'* on the opposite side of AB; after which extend lines from the points *r'l'k'* parallel to AB. This gives the point k at the same distance from AB as the points *a* and *a'* and the line *ll'* at the same distance as *cc'*. The rest of the same group of parallel lines are found to be similarly situated with repect to AB.

KERFING

When any curved part is to be covered with a board or molding, difficulty would be met in making the wood bend unless it was a thin veneer, hence a number of saw kerfs are made in it, so that it will bend easily.

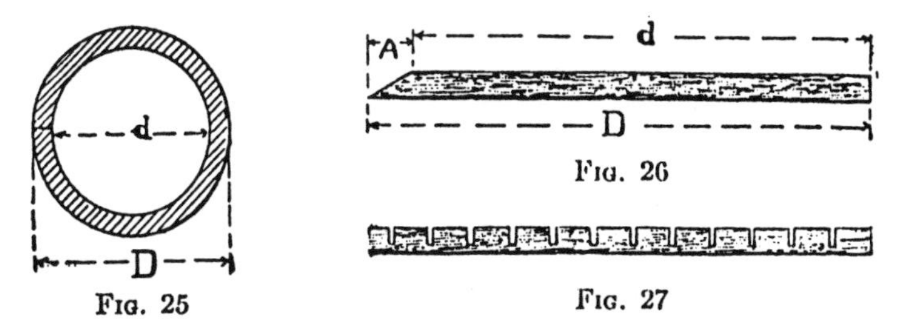

Fig. 25

Fig. 26

Fig. 27

Suppose that the most simplest example is taken to bend a flat board around a circle; the board when bent will be in the form of a circle, as shown by the shaded part of Fig. 25. If the board is thicker than an ordinary veneer, it will not bend to a circle without breaking—tearing apart on the outside and crushing together on the inside. The reason for this is that the inner diameter is smaller than the outer diameter. The same is true of the circumferences; the inner circumference is smaller than the outer circumference; therefore the problem is to make the inner edge of the board smaller than the outer, assuming that to be the required size for the circumference. Now if the inner edge *d*, Fig. 26, was shortened simply by cutting off the end as shown, every section would be the right length to go around in concentric circles, but as the fibres of the wood are unable to move, this method will not do.

Suppose, instead, that along the inner edge of the board we cut out little pieces at even distances, so that the sum of these distances would be equal to A, which is the amount that the inner circumference must be shorter than the outer circumference; then we would practically have the same result as if the wood were made of layers of fibres that would slide on each other as the board was being bent.

Fig. 28

The little pieces that are cut out of the inside edge of the board are in practice merely the material that is removed by a saw kerf; hence the reason why this method of bending a board, molding, etc., is called "saw kerfing," or simply "kerfing." Fig. 27 shows how a board looks when ready for bending. To get good results it is, of course, necessary that all kerfs should be the same; hence only one saw must be used, and it must cut to the same depth at even distances apart. For ordinary thickness of material, such as used for interior trim, the kerfs are made to about $\frac{1}{8}$ inch of the outer edge. This gives the outer edge a thickness that makes it practically a veneer, which, as we already have mentioned, is readily bent.

Knowing the width of the saw cut, the thickness of the board, and the radius or diameter, we may calculate the distance apart and number of saw kerfs there should be, but this would be a waste of time, as we can adopt a simple, practical method that will give us this information without much trouble.

Take a board the same thickness as the piece to be bent and make a saw kerf in it of the required depth and at a distance from one end equal to the radius of the required curvature, as shown in Fig. 28. Then bend the board until the kerf is closed; the distance AB, Fig. 29, through which the edge

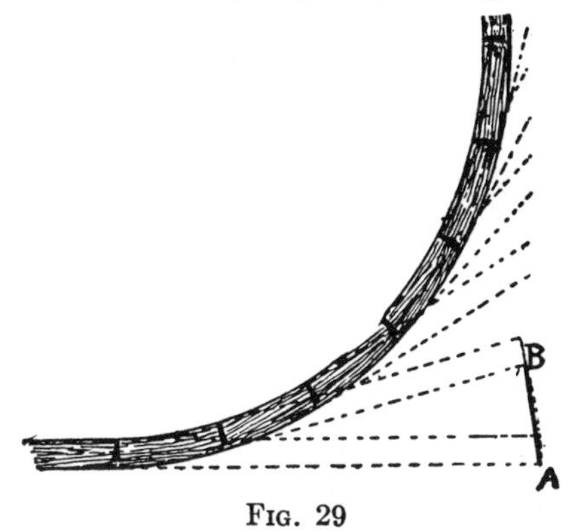

Fig. 29

of the board moves, is the distance between the saw kerfs. Readers who study Fig. 29 will understand why this method is correct.

The distance AB is the chord of the arc of circle the board is bent to, made by the one saw kerf. If more saw kerfs are made, the board will bend up as illustrated, and this clearly shows that the distance between the saw kerfs is the chord of the arc, and equal to the distance AB, as explained above.

If the board to be bent to a curve is on a rake, make the saw kerfs so that they will be vertical when placed in position.

BRACKETS FOR COVES OR MOLDINGS

It is a simple job to lay out the brackets for a cove; the only difficulty will be at the angles; these brackets must of course be longer. It is a good plan to have them of thicker wood than the common brackets, so that there will be room enough to properly nail the laths.

Let Fig. 30 be the shape of the common cove brackets. Divide the width AB into any convenient number of equal parts, and draw lines parallel with its side AC.

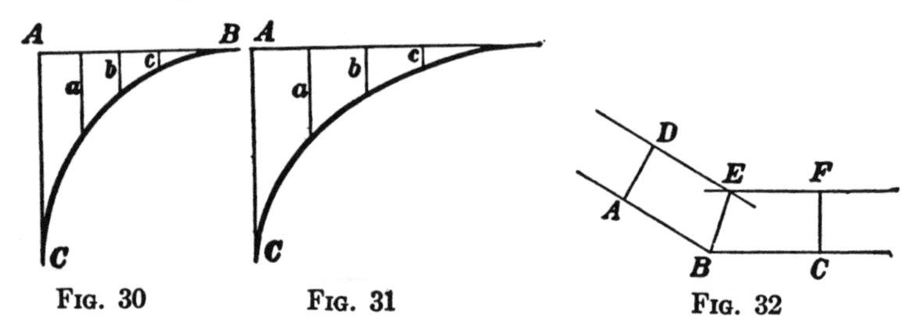

FIG. 30 FIG. 31 FIG. 32

After the proper width of the angle bracket is found, divide this width into the same number of equal parts as the common bracket, and draw lines parallel to its side, AC, Fig. 31, which is, of course, the same depth as the common bracket. Make the depth of each line, *a*, *b*, *c*, etc., the same depth as the corresponding line in Fig. 30. Through the ends of these lines draw a curve, and this will give the correct form for the hip bracket.

The above method is correct for all sizes and shapes.

The method of finding the width of the hip cove bracket is very simple:

For a square corner, the width of the hip bracket is the

diagonal of a square whose sides are equal to the width of the common cove bracket.

For other shapes—pentagon, hexagon, octagon, etc.—draw two lines, making the angle formed by the hexagon, etc., as shown in Fig. 32 at AB and BC. On these lines erect perpendicular lines AD and CF; make the length of these lines equal to the width of the common cove brackets, and at the extremities of these lines draw lines DE and EF parallel to AB and BC.

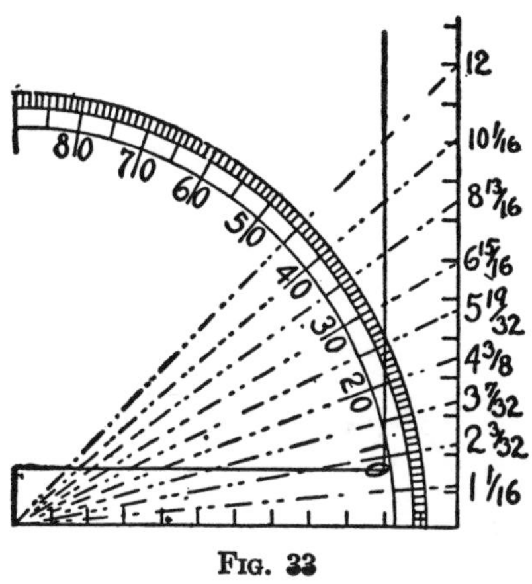

Fig. 33

From the point E, where these two lines intersect, draw the mitre line BE, which will be the required width of the hip cove bracket.

The above method can be applied to get the size of hip brackets for both single-curved and double-curved sections, and will give the correct form, no matter how large the brackets may be.

THE STEEL SQUARE

Information about the use of and descriptions of the marks on carpenters' steel squares are given in quite a number of books. The three following figures are taken from D. L. Stoddard's

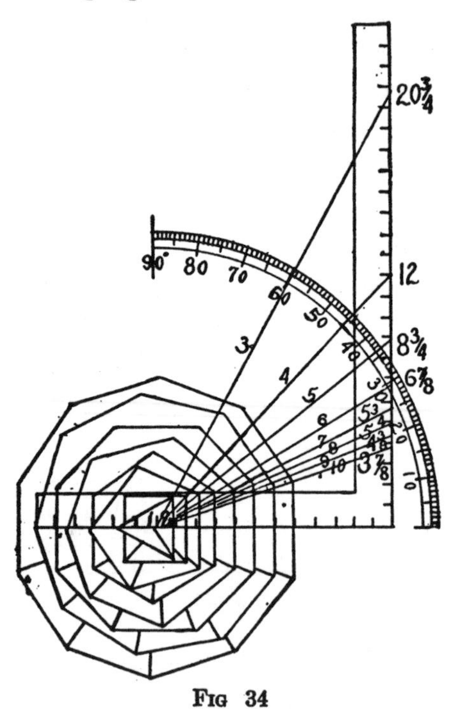

FIG 34

Steel Square Pocket Book (which is published by the publishers of this book, at fifty cents).

Fig. 33 shows how to get various angles by the use of the square using 12 on the tongue and the number on the blade as given on the right of the drawing. For angles beyond 45°, reverse the square.

To lay out any polygon, it is only necessary to know the angle to get the marks on the square. Fig. 34 shows it all up to a ten-sided figure.

There is sometimes a little misunderstanding about pitches.

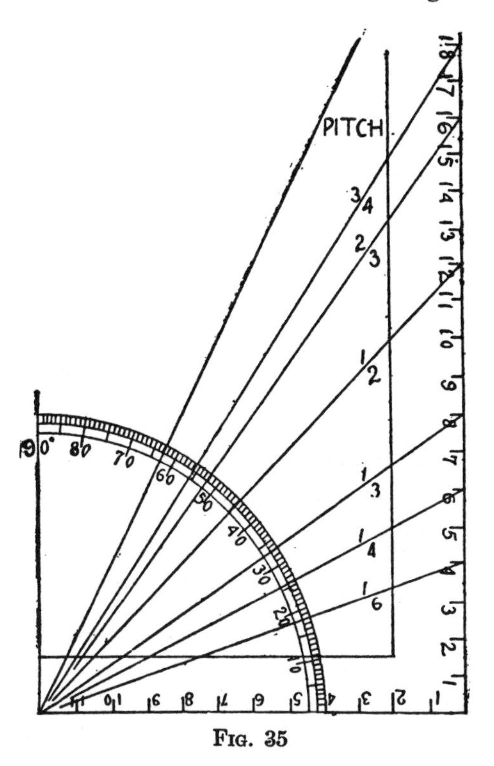

FIG. 35

Fig. 35 shows the pitches most in use and the marks to take on the square to get the cuts for the rafters. More information on the use of the square for laying out rafters will be found in the section devoted to roof framing.

THE TWO-FOOT RULE

The following article is by Charles G. Peker, and will be new to most mechanics:

Many special tools are used by workmen for laying out work, but sometimes these may not be at hand; every mechanic, however, will have his two-foot rule.

Most people think it can only be used for measuring length, but it is capable of quite a number of applications that may be frequently used when no other device is at hand.

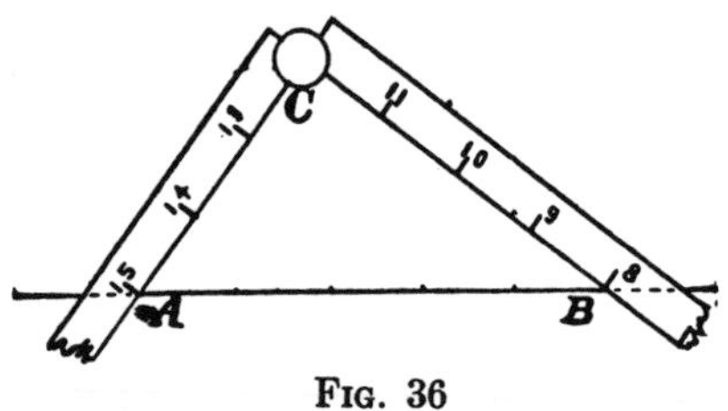

FIG. 36

TO DRAW A STRAIGHT LINE

This needs no explanation, except to say that to draw a line longer than the rule, move the rule along and continue the line.

TO DRAW A RIGHT ANGLE

Draw a line of any convenient length, as AB, Fig. 36, and on this set off a distance of 5 inches; then lay the rule as shown in the figure, with one fold touching B at a mark 4 inches away from the joint, and the other fold touching A at a mark 3 inches away from the joint. Then the angle ACB will be a right angle.

TO DRAW A PERPENDICULAR

Another method on the same principle is used when it is

desired to draw a line at right angles to another line, that is, perpendicular to it.

Draw the line AB, Fig. 37, and from the point where the perpendicular is to be erected, set off 4 inches; then lay the

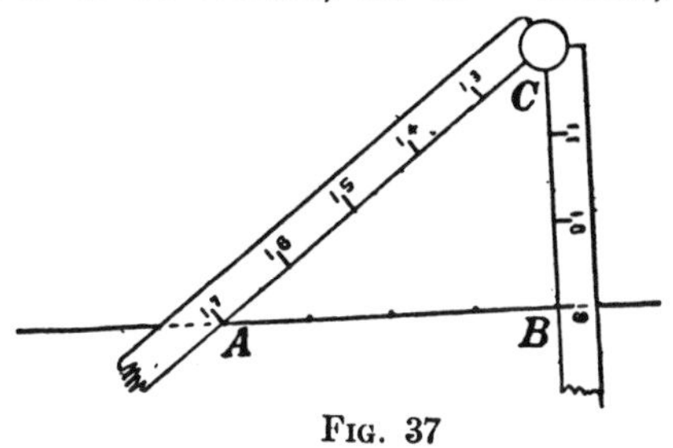

FIG. 37

rule as shown, so that the inner edges of the rule touch the points A and B; the distance from A to C is 5 inches, and the distance CB is 3 inches. A line drawn along the edge BC of the rule will be perpendicular to AB.

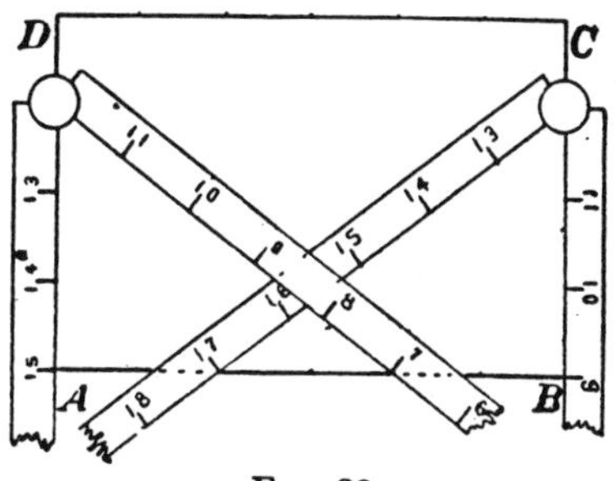

FIG. 38

TO DESCRIBE A RECTANGLE

To do this it is necessary to draw four right angles, this being done by the method just described.

To show just how this is done, an example will be worked out.

To draw a rectangle 6 by 4 inches, first draw a line, AB, 6 inches long, Fig. 38; then at A set the rule and draw a perpendicular, AD, and make it 4 inches long; at B repeat the method, drawing the perpendicular CB; this should also be 4 inches long. Joining the points D and C will complete the rectangle.

It is not necessary to apply the rule at the points D and C to get the line DC perpendicular to AD or BC, as when DA and

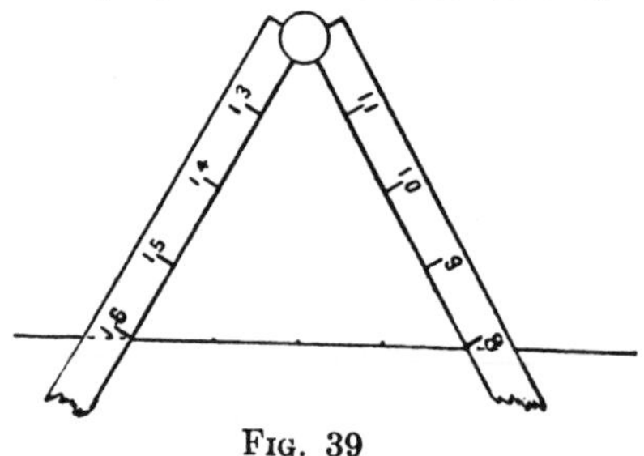

FIG. 39

CB are the same length and perpendicular to AB, DC will be parallel to AB and consequently perpendicular to AD and CB.

TO DRAW AN ANGLE OF 60°

Draw a straight line, and on this lay off a number of inches, say 4; then place the rule as shown in Fig. 39, with marks on each leg of the rule the same distance from the joint as are laid off on the line (4 in this case), touching each end of the given line. Any two lines of the equilateral triangle thus formed make an angle of 60°.

TO DESCRIBE A HEXAGON

A hexagon may be easily described by repeating the method

of drawing an angle of 60°. Suppose it is desired to draw a hexagon with sides 3 inches long.

Let AB, Fig. 40, be one side of the hexagon, 3 inches long; extend it 3 inches longer to C, then place the rule with marks 3 inches from the joint at B and C; draw the line BD and extend it 3 inches to E; repeat the same method to get the line DF, etc., and thus obtain the hexagon ABDFGH.

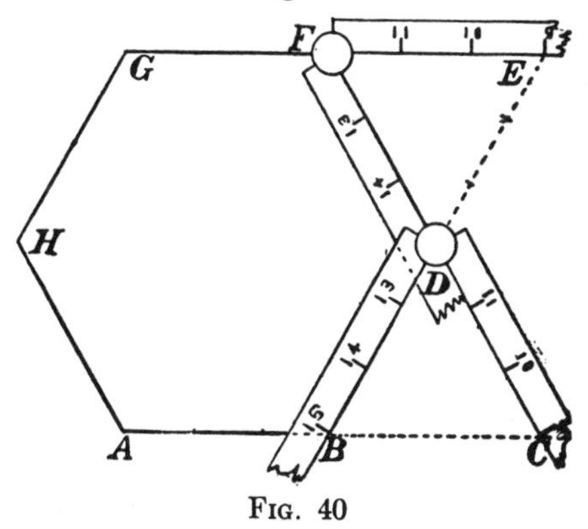

Fig. 40

Other problems could be given, but trust that enough have been given to show that the two-foot rule is more useful than most mechanics think.

GLUE AND ITS USE

Good glue is hard, clear (not necessarily light-colored, however), and free from a bad taste or smell. Glue which is easily dissolved in cold water is not strong. Good glue merely swells in cold water, and must be heated to the boiling-point before it will dissolve thoroughly, says John Phin, the author of *The Workshop Companion, Cements and Glue*, etc.

Glue being an animal substance, it must be kept sweet; to do this it is necessary to keep it cool after it is once dissolved and not in use. In all cases keep the glue-kettle clean and sweet by cleansing it often.

Good glue requires more water than poor, consequently you cannot dissolve six pounds of good glue in the same quantity of water you can six pounds of poor. The best glue will require from one-half to more than double the water that is required with poor glue, which is clear and red, and the quality of which can be discovered by breaking a piece. If good it will break hard and tough, and when broken will be irregular on the broken edge. If poor it will break comparatively easy, leaving a smooth, straight edge.

In dissolving glue it is best to weigh the glue and weigh or measure the water. If not done there is a liability of getting more glue than the water can properly dissolve. It is a good plan, when once the quantity of water that any sample of glue will take up has been ascertained, to put the glue and water together at least six hours before heat is applied, and if it is not soft enough then, let it remain longer in soak, for there is no danger of good glue remaining in pure water, even for forty-eight hours.

The advantage of frozen glue is that it can be made up at once, on account of its being so porous. Frozen glue of same grade is as strong as if dried.

If glue is of first-rate quality, it can be used on most kinds of woodwork very thin, and make the joint as strong as the original. White glue is only made white by bleaching.

In blind nailing (which will be explained further on) the

slivers of wood have to be glued down. To do this properly, the glue ought to be good and well prepared, and this rule ought to apply to all work.

In selecting glue we have had the best results with the old-fashioned cake glue. It requires a little more time to prepare, but does not call for any more attention or work on the part of the person that prepares it.

See that it is clear and transparent, and that it has no putrid smell when soaked in water. Cook it thoroughly, but be careful that it does not get burnt.

As you cannot rub together the two pieces of wood to be joined, rub the glue well in with a stiff brush; hold in place either with a vise or some kind of a heavy weight, and allow plenty of time to dry.

The fibres of the wood will then tear apart rather than separate from the glue. The error made by most people is that they hurry the preparation and the drying.

WORKING HARDWOOD

The following practical hints on this subject have been contributed by Owen B. Maginnis, author of *Roof Framing Made Easy* and *How to Measure Up Woodwork for Buildings:*

The following information may prove useful to those carpenters who have never worked hardwood, and who may be suddenly given a job to do. When nailing never drive cut-iron nails without first boring for them. A German bit is the best to use; they cut clean and easy, and pull out the cores. Wire nails, however, are best for the work, and if the wood be very hard, as heart ash or quartered oak, the point should be dipped in wax or soap. A very convenient grease box can be made by boring a $\frac{3}{8}$-inch hole in the haft of a hammer and filling it up with wax or soap; it is always at hand and not in the way. When nailing in panel molding, never drive the nail slanting, so that it will go into the panel, but keep it nearly parallel with the face of the panel, so that the nail will enter the edge of the stile rail or muntin. By driving it thus, it will hold the molding on the frame, but if it goes into the panel, when the panel shrinks the molding will draw away from the edge and likely split the panel.

When possible nail all work so that the nail-holes will cover; for instance, when setting jambs, nail through the edges into the studding, as polished or varnished work showing the natural wood is marred by nailing through the face. If it must be nailed on the face as trimming, be very careful not to split the work, and use a small-pointed set. Be sure there is no grease left on the face of the hammer after each nail is driven, because it is liable to bend the next nail. The best way to make inside joints with hardwood is to scribe them. Often the stuff is warped, and cutting a square joint only means fitting, which can be avoided by scribing the joint. If it is compulsory to nail back a piece of very much twisted stuff, like hazel or sycamore, the mechanic should be very cautious, for kiln-dried timber is so brittle and non-elastic that it will not readily yield to strain and will crack like glass. The best plane to use on all hardwood is of course the iron plane, especially where the wood is curly or cross-grained. The Bailey plane is about the best for general work, and keep it sharp and keen. Dull tools may do fairly well on pine, but on hardwood they are useless, so keen edges are indispensable for clean work, so that little scraping need be done. A good iron block plane is also necessary. I would recommend that all cuts and mitres be made direct from the saw. A good workman will make his calculations carefully, and insure the certainty of his cut before making it.

Finally, proceed carefully and steadily, and don't bang the work with the hammer. Use a block if you must drive the stuff, so as not to mark it, and make a good, clean job.

BLIND NAILING

In places where wood is to be stained and polished, as is usually the case where any of the hardwoods are used, it is not

FIG. 41

desirable to have the nail- or screw-heads to show. One method is to drill a shallow depression in the wood, and, after the screw or nail is driven in, to glue a circular piece of veneer in the depression.

The favorite way, however, is to raise a "chip" or "sliver" with a chisel or gouge, then drive in the nail or screw and glue down the sliver.

In Fig. 41 the wood is raised by means of a firmer chisel. A sharp knife should be employed to draw lengthwise with the grain two deep cuts the width of the chisel, as this will prevent the sides from splitting. The chisel should be set at a steep angle at first till the proper depth is reached, and then made to turn out a cut of even thickness until there is room to drive in the nail or screw. If too sharp a curve is given, the sliver is likely to break apart in being straightened out again.

A useful aid in doing this kind of work is the chisel gauge shown in Fig. 42, the use of which insures a sliver of the proper length being made each time.

Fig. 43 shows how a sliver is raised by using a gouge. To do this nicely, a gouge about three-quarters of an inch across the face should be used, and the curve should be quick. In this case no knife cut is needed, as the corners of the gouge will cut as it progresses.

FIG. 42

The cut being made and the sliver slightly raised as shown, the screw or nail may be driven without disturbing either the sliver or cut underneath.

See that the head of either screw or nail be sunk beneath the surface of the recess, so that the sliver will fit back in its place without obstruction.

Now take properly prepared glue and, after warming the sliver and recess with a warm cloth, cover the under side of the sliver and the wood underneath. Press down the sliver in place, then rub with the face of a hammer until the glue holds. When dry the whole may be dressed off and finished. The glue must not be too thick.

Another way is to glue the sliver down and then take a flat piece of pine about an inch thick and glue over the sliver, rubbing the pine block back and forth until the glue holds.

The pine block is left on until dry, when it may be easily split off and the wood cleaned and finished.

SETTING DOOR JAMBS

The hanging of a door depends on how the jambs are set. Mr. J. E. Nelson gives the following useful information on this subject:

Fig. 43

In order to set jambs right, a partition studding should be set plumb and special care taken to get straight studding next to the door jambs. They should be straight both ways and set as near plumb as possible. If they are out of plumb sidewise, it will cause the door to scrape or touch the floor when opened, if hung on one side; or if hung on the other side, it will cause the door to rise as much from the floor.

I leave an opening 2′ 10¼″×6′ 10½″ for a 2′ 8″×6′ 8″ door; when I figure on using ⅞-inch stuff for jambs, gaining side jambs ⅜-inch for head jamb, and that leaves ½-inch to plumb and wedge or furr out on. I always aim to have both door studs set plumb, so that the door may be hung on any side that may be preferred.

I cut my side jambs $\frac{3}{8}$ inch longer than they should be, nail head jamb to them, and stand them up where they belong; then wedge at the top so as to make them stay in place; then see if head jamb is level or not.

If the head jamb is level, scribe both sides $\frac{3}{8}$ inch, but if not level, scribe the side that is low $\frac{3}{8}$ inch, and the side that is higher scribe a little more than the $\frac{3}{8}$ inch, so as to make head jamb level when both side jambs are cut off.

After taking down and cutting off, the jambs are replaced, nailing the one jamb up tight to the studding; this is the side the door is to be hung on. Then cut a board the same width as the head jamb, to place on the floor between the side jambs. This board must be cut square at both ends, and, for the door mentioned above, should be 2′ 8″ long. Now wedge the bottom of the other side jamb tight to it, then wedge and nail head jamb in place, first, however, plumbing sidewise. Next, I use a straight-edge and wedge side jamb up to it; also look at the other jamb, and if not straight, wedge it up until it is O. K., by using the straight-edge against it. When I have set jambs in this manner, I do not smell any sulphur or brimstone when the door hangers come along.

Some are also in the habit of furring out with lath on the inside of studding where the jambs are to be set, and placing the lath at a guess for the mason to plaster to. That, I think, makes a poor job, as plastering will be thicker in some places than others. The way I do is to select lath of as even thickness as I can, and nail them on top of the lath all around the door studding, head trimmer included; then I can make my jambs the same width all the way through, and it will leave a smooth face to case on.

These remarks are principally intended for beginners in carpentry work, as older hands know from experience how they should be placed; but I have seen contractors put their poorest men setting partitions, and I have often wished they had to set the jambs and hang the doors themselves.

HANGING DOORS

The following information on hanging doors is contributed by Mr. H. J. Aurlie, the author of *Rafter and Brace Tables:*

In the first place the butts on common doors should never be secured where the screws will enter tenons of the cross stiles. The bottom stile will average 10 inches on ordinary sized doors, but on very large doors they may be up to 12 inches in width, and the top stile will average 5 inches, or the same width as the side stiles. The rule adopted by most good carpenters throughout this part of the country is to place the lower butt a half inch above the tenon of the lower cross stile, and the top butt 6 inches from the top of the door, while there are some that place it 7 inches from the top. Either is right, but, personally, I prefer 6 inches; and the distance, to my way of thinking, should never be over 7 inches.

If there are three butts to the door, the middle butt should be placed half way down between the other two, though in some special cases, on account of arrangement of panels, this rule is sometimes varied.

The lock in common doors should be 2 feet 11 inches from

floor, though sometimes this distance may have to be made higher or lower to get away from a tenon.

In ordinary five-panel doors the lock comes in the heart; that is where the cross panel comes; this leaves a space about the right height for the lock between the two tenons of the cross stiles above and below this cross panel.

It is decidedly poor policy to have to mortise out most or part of the tenon of the cross stiles, but this will often have to be done in four-panel doors to get the height of the door knob the right distance; and I might say in this connection that it would be a good idea if the manufacturers of doors would pay some attention to where a lock might be placed.

A word about easing doors. In nailing on the casings, they should be placed back far enough so as not to interfere with the butts—about ⅜ inch will be sufficient. It makes a bad job where the casings are so close that when the butts are fitted in the jamb, the tips or balls of the loose pins will have to be dug into the casing edge; and once in place, the pins cannot be got out unless the butt is loosened or the casing gouged out.

FITTING DOORS

There is quite a nack in properly fitting a door; in fact, it is half the hanging. Mr. L. M. Hodge describes a good and short-cut method for doing it:

I will not attempt to give any specific directions as to how a door should be put up, but rather describe a method of accomplishing the work, which I have found to be a time-sâver and also to insure accuracy; and while this method may be old and well known in some parts of the country, still it may be new to some of the younger "chips," who will find the principle as applicable to the fitting of windows as well as to hanging doors.

To mark the location of the hinges on a door or jamb, take a thin flat rod, a few inches shorter than the door, and mark on

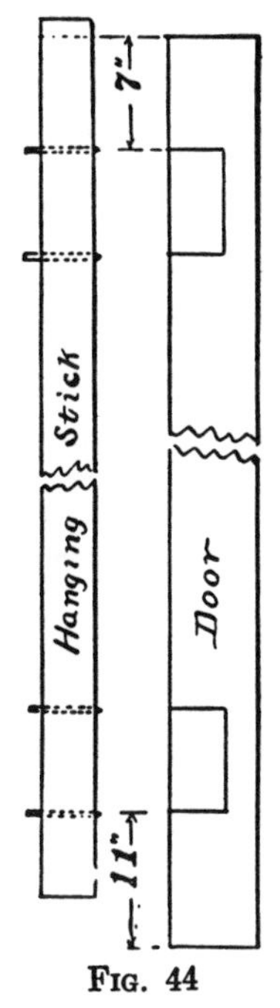

FIG. 44

it the location of the hinges as desired, and drive small wire brads through at these points, just letting them project through enough to make a slight scratch; sharpen them nicely, then make a mark on the upper end of the rod down about 1/32 of an inch from the end, or the amount of clearance that you wish the door to have at the top, and you are ready for operations. Fig. 44 shows the idea.

To use the instrument on the jamb of the door frame to which you wish to hang the door, place the top of the rod against the head jamb in the rabbet for the door and press it against the jamb, and the sharp points will prick off at once the exact location of the hinges; then laying it on the hinge edge of the door, with a mark A (which represents the desired clearance or space left between the top of door and head jamb) at the top of the door, in like manner prick off the location of the hinges

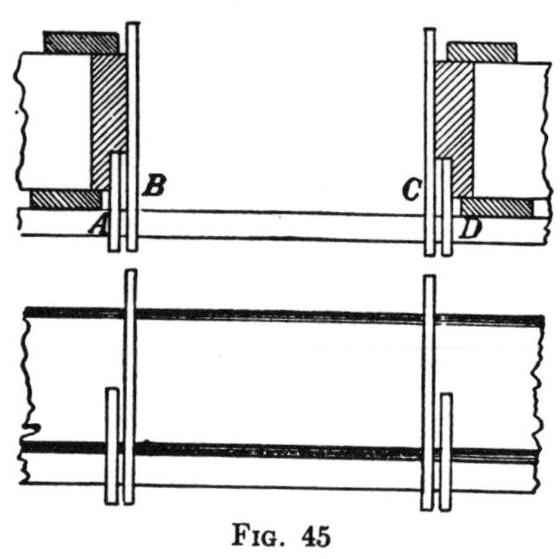

FIG. 45

on the door, and if the hinges are accurately let into the door and jamb at these points, the hinges will slip together easily and work perfectly free; thus it will be seen that by this method the hinges may be put on both jamb and door without setting the door into the frame in order to mark for the hinges, which saves no little amount of work and time to the operator.

An easy way to mark in thresholds is described by Mr. F. A. Williams:

To mark in thresholds when they are square, twisted, or

splayed, I place a straight-edge 8 or 9 inches by $\frac{3}{4}$ inch thick across the casings, as per sketch in Fig. 45. Then place narrow straight-edges, A, B, C, D, fair with the door frame and rebate and mark across the 9-inch straight-edge, or they may be tacked to the straight-edge.

Next place the threshold and the 9-inch straight-edge together, as shown in the lower sketch, and transfer the marks from straight-edge to threshold. In this way the threshold will be certain to be a good fit. Mark with a sharp knife and cut a little long.

FITTING WINDOWS

Mr. L. M. Hodge further says that in fitting windows, take a flat rod similar to that shown in Fig. 44, the width of which equals the thickness of the sash, and after fitting an upper sash and fastening it in its correct position in the frame, cut a bevel on one end of the rod to fit the pitch of the sill of the window frame, and, standing it on the sill in the run of the lower sash, mark the rod at the top, or 1/32 of an inch below the top of the meeting rail of the top sash; this gives the exact length of the lower sash on the rod, together with its lower end bevel to fit the sill; then all we have to do to accurately fit all the sash of that size is to place the rod flatly on each side of the lower sash with the square or top end even with the top of the sash, and mark along the beveled end of same; then, by having one edge of the rod straight, it can be laid on the outside of the lower rail intersecting the beveling marks on the sash and the mark made across the face of the rail.

I might say that as the width of ordinary windows seldom exceeds the length of the lower sash, the exact width of the sash can be marked on the rod (allowing of course the desired clearance); thus all the windows of that size, including both upper and lower sash, may be fitted at a bench and carried and set in their places in the frames.

I might further say that it is well, when using the above method, to fit the longest windows first, then the same rod may be cut off at the top or square end to fit the shorter ones by.

A question often asked is "how to cut out pockets for windows with weights." Fig. 46 shows clearly the method of cutting the pulley stile of the window frame at a bevel; A is a front

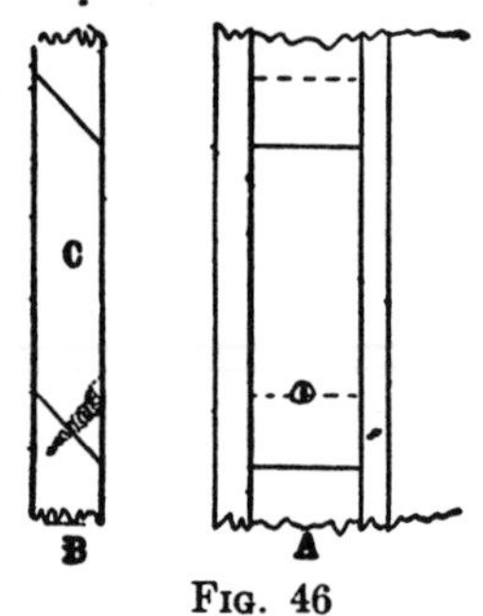

FIG. 46

view and B a section of the stile. By removing the screw, the piece C may be taken out and the weights and sash cords repaired without removing the stile.

NOTES ON FRAMING

The whole of the timber used for framing to be the best of its kind, sawn die square, well seasoned, and free from shakes and other imperfections impairing its durability and strength.

All joists to be placed with crowning edge upward, and stiffly and fully spiked at each end to bearings, and to each other when they come together. Joists should be cross-bridged every 6 feet, with 2″×4″ stuff properly cut in between timbers, as soon as the joists are leveled and secured.

Headers and trimmers should be 3 inches thick, properly framed and spiked together, leaving all openings of sufficient size for the finish of stairs, chimneys, etc. In no case should the wood come within 2 inches of the brick around any smoke flue.

Studs should be set plumb and securely spiked to sill. They should be spaced 16 inches on centres, the same as the joists.

In a balloon frame the sheathing should always run diagonally, so that it braces the frame.

Use plenty of nails and stagger them; do not have two nails cutting the same wooden fibres.

Pay particular care to have solid corners both on exterior and interior angles.

Take particular care to get the sill square and true, as the entire frame work depends on it.

SIDING

The question of siding the frame is an interesting one, and Mr. F. W. Pape gives the following good hints on this subject:

I have found that the most trouble with young carpenters is in learning to weather board properly. At least that has been my experience in hiring hands. They are not careful enough about running the siding straight and making it come up to the window sills and up over the door and window caps. Thinking it will interest your readers, I will give the methods I follow.

Fig. 47 is a gauge for the siding, and this gauge is used throughout the job. The first point I see to is that the siding is started straight with the lower edge of the sill or wall, as the case may be.

Fig. 48 is a gauge which I use in making the ends at corner boards and frames. This is done with a carpenter's pencil sharpened to a chisel point.

Take a slat about 3 feet long and mark on it a number of lines, the distance between them to be the same as the distance

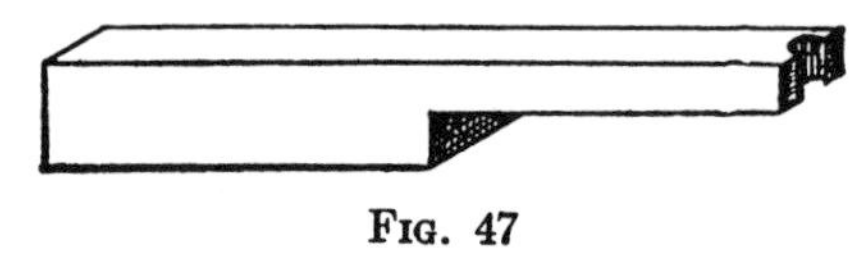

FIG. 47

the siding is to go to the weather. Hold this measuring stick up under the window sill, and as the stick is numbered, I can tell at a glance how many more boards it will take to run up to the window and whether I will have to set my guide nails a little above or below the gauge line to make the siding come even with the bottom of the window sill. The same method is used on other parts, to the top of window or door caps, etc.

FIG. 48

Now for heading joints. I saw the first piece of siding a little under, and do not nail the end until I have the next piece of siding sawn and fitted. This is done by slipping the second piece under the first, and marking with a sharp pencil. In this way I am sure to have a fit. Readers who use these rules will save time and will have but little trouble to make a good job.

Mr. E. W. Reynolds says that when the frame is complete,

commence siding on ends of building, sawing, after nailing on, smooth with corner posts; then put on sides, sawing smooth with outer part of siding on ends. When siding is complete, take two planks, any thickness desired, not less than 6 inches in width, and make a trough or comb board and place on corner and nail

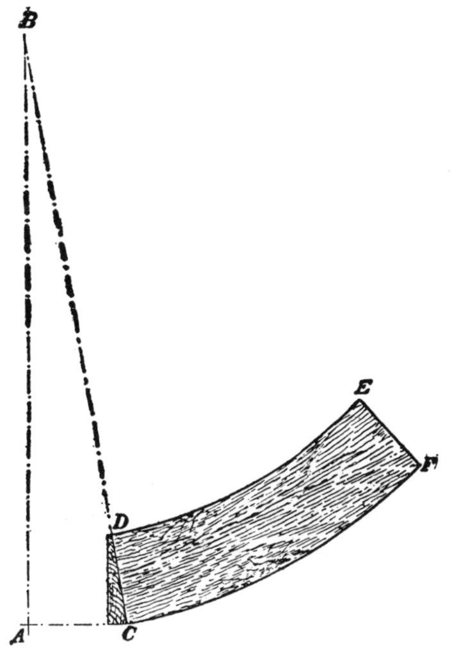

FIG. 49

through siding into corner post. This plan has three advantages: (1) It excludes water from corner posts. (2) Can be removed and new ones put on easily. (3) Saves time, as there are no joints to make at corners.

To place siding on a circular tower or corner of a building, it is necessary that the siding be of a certain special shape.

First make a drawing of a section of the siding as shown by the shaded portion of the sketch from C to D, Fig. 49.

Draw a horizontal line at the bottom of the section of siding, making the distance AC equal to the radius of the curved part. Draw BC tangent with the outside of the clap-board section. The line BC cuts the centre AB at the point B.

From B as a centre, and with a radius BC, describe the arc CF; with the same centre describe also the arc DE. The portion DECF is the correct shape of the siding necessary to secure level lines running around the circular part.

The curve as found above is marked on a piece of board or cardboard which is cut out and used as a pattern for the siding, each strip of which is marked and worked out with a draw knife and plane.

The length of the pieces should be such that they will end at a stud, so that they can be properly nailed.

SHINGLING

Shingles come in bundles of 250, and are usually 16 or 18 inches long, but they vary in width, the average width being 4 inches; hence 1 bundle will lay one course 1,000 inches long.

To cover 100 square feet of roof will take

900 shingles if laid 4 ins. to the weather.
800 shingles if laid 4½ ins. to the weather.
720 shingles if laid 5 ins. to the weather.
655 shingles if laid 5½ ins. to the weather.
600 shingles if laid 6 ins. to the weather.

Where shingles have to be cut, as in hip roofs, add 5% to the above for the waste.

About 6 lbs. of fourpenny nails are needed to lay 1,000 shingles. If the nails are moistened with saliva before being driven, it will prevent the shingle from splitting.

To insure water-tight joints at hips, valleys, ridges, etc., tin is used at these points, and this is called "flashing."

Fig. 50 shows a portion of hip roof. It will be noticed that at the hip there is a continuous row of joints; this would allow water to work its way into the interior. To prevent this, flashings are used, that is, pieces of bent tin are placed over the hip, and the shingles cover this. These pieces of tin are placed over

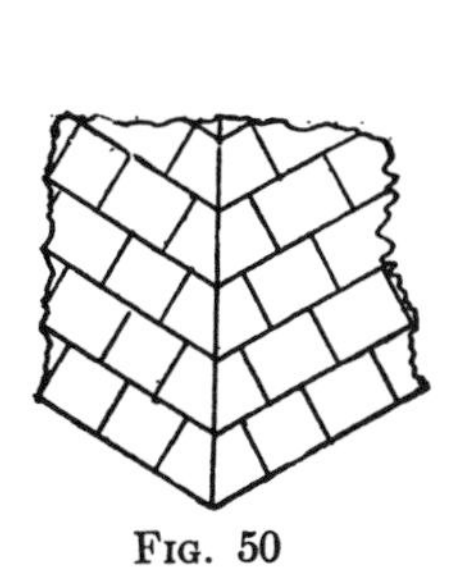

Fig. 50

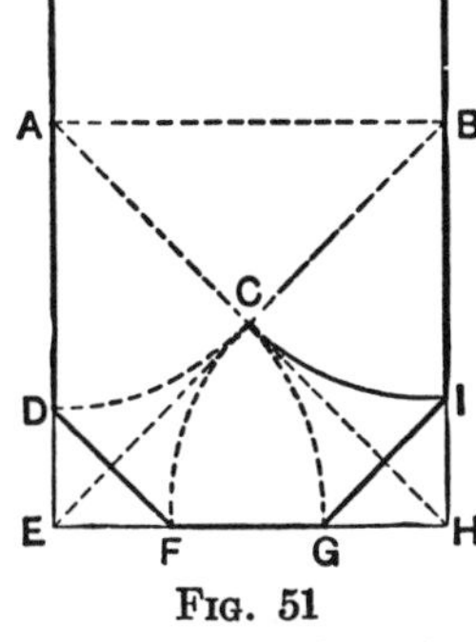

Fig. 51

each course of shingles the same as a shingle would be laid if it were possible to bend it over the hip.

To lay out an octagonal shingle, draw a line AB across the shingle, as shown in Fig. 51; this line to be parallel to the butt the distance AE or BH is equal to the width of the shingle. This makes ABEH a square. Draw the two diagonals AH and BE; these intersect at the point C. Now with a radius equal to EC, and the points AEHB as centre, describe the arcs. Joining the points D and F and G and I completes the marking, and the shingle is now ready to have triangles DEF and GHI cut off.

A short cut in getting the lengths of the beveled shingles in a gable is contributed by Mr. F. C. Bell:

In Fig. 52, after we lay our course up to the roof as far as we can without cutting, we reverse the shingle, as the one marked D, allowing the scollop A to touch the planceer board, and marking the other side of the shingle at B, which will be the correct length of the short side of the shingle, as will be seen by comparing with C. We now lay on the bevel and have our

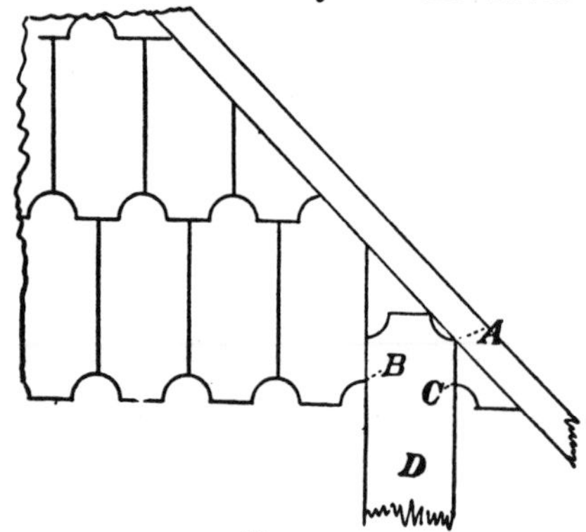

FIG. 52

shingle laid out, without measuring with the rule, accurate and in half the time.

FRAMING A FLOOR WITH SHORT TIMBERS

A problem frequently asked is "how to frame the joists for a floor the dimensions of which are larger than any available lengths of joists, and to have no central support."

A method of doing this, say for a floor 35 feet square when the largest joists to be had are only 25 feet long, is shown in Fig. 53,

Another method is shown in Fig. 54. AA are headers, made by putting two joists together. These are mortised and tenoned together. The other beams are tenoned into the headers, or may be held by any of the various iron joist hangers on the market.

BUILDING UP A BEAM

In joining several timbers together to get a deeper beam, a good method is that shown in Fig. 55. Two or more pieces of timber are made into one compound beam by nailing boards diagonally on each side in opposite directions.

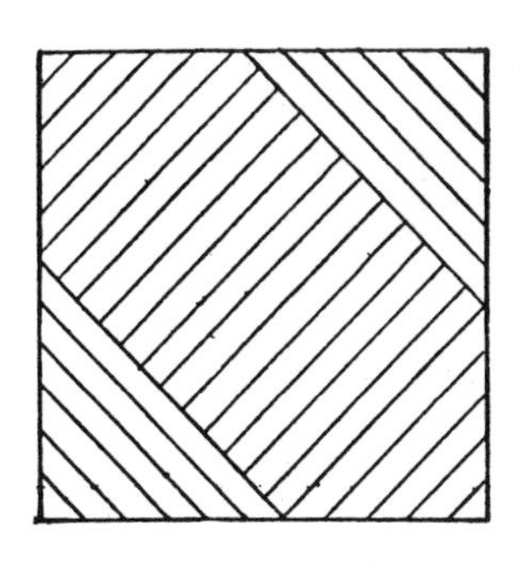

Fig. 53

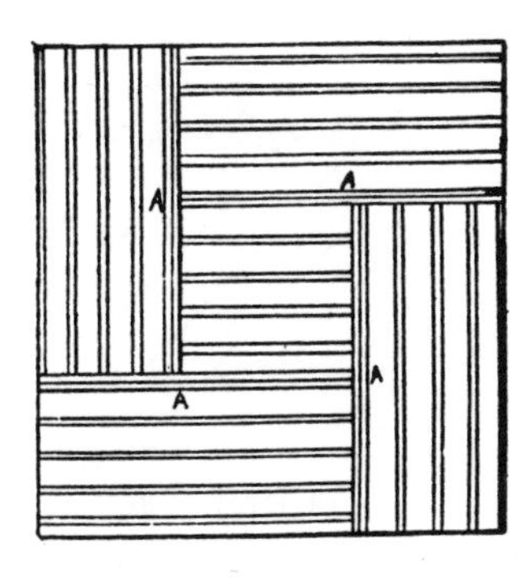

Fig. 54

LAYING FLOORS

As a general thing floor laying is considered a very easy thing to do, and it is a common expression, that "most anybody can lay a floor," but it is a very mistaken idea. It is true that this "most anybody" is put to floor laying, but the results are anything but satisfactory except to those who believe in doing their work with cheap and irresponsible help. The following notes are by Mr. Jonathan Torry:

Nothing adds more to the looks of rooms, whether in the parlor, kitchen, hall, or in a work-room, than a smoothly laid tight floor. The fact that a room is to be covered with a carpet is no kind of reason why the floor should not be well and smoothly laid. The edges of boards in unevenly laid floors are very destructive to any kind of carpets, and oilcloths in kitchen, pantry, or halls are necessarily very short-lived if laid over bad floors, and the musical floor so often found, especially in houses costing from $2,000 to $4,000, is detestable, more especially to those persons who are extremely nervous and are frequently hearing strange sounds.

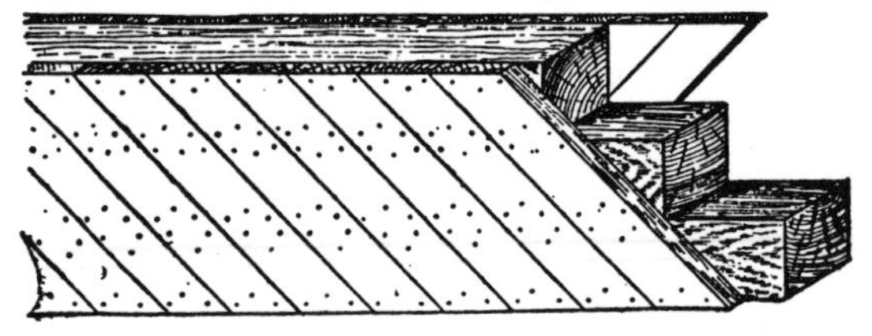

FIG. 55

The idea of simply getting something to walk on to keep one from falling into the room below is not a good one. Good floors add very materially to the strength of a house, and the more pains taken to *lay* and *nail* a floor well, the better the structure. The common way of driving a floor together with hammer or hatchet or maul, directly on the board being laid, should never be allowed, for in very many, if not in all cases, the tongue is broken off somewhere, and as soon as the floor shrinks a little, a large opening is found, and it is not uncommon to find floors full of holes from this cause. Not one piece of flooring should be laid by driving up the board. If no clamps are used, and it is being blind nailed, a good-sized piece of matched stuff ought always to be used, and through and through nailing, a parallel strip ought to be nailed down firmly to spring the floor close. Blind nailing needs special attention, and no blind-nailed floor should have the stuff worked more than 3 inches wide.

The great trouble with blind nailing is that the nails are driven too flat. The angle should never be more than 45° and less than that should be the rule with good nails. The use of steel nails makes good floor laying more easily done, because we get a stronger nail with less size. One great fault with floor laying is that the layer allows the floor to gain at one end, or, if a long floor, draws it up uneven and makes it crooked, and a crooked floor once started is a hard thing to straighten.

I notice many objections to blind nailing, but I think the fault is oftener in the conditions than with the nailing. Either the flooring is too wide or the layer is too careless in laying it down, driving in his nails carelessly and at any angle but the right one. In laying *any* floor much care should be taken not to split off the tongue in nailing. It is usual to drive the nail home with the hammer, but unless the driver is an expert with the hammer, the odds are that where every nail is, there will be a nick in the edge of the board.

With nice floors the holes for the nails should be drilled or bored with some tool that will be sure to not split off the tongue. Much pains should be taken to start a floor straight to begin with, and as much pains should be taken that the floor is laid evenly and not allowed to crawl and become crooked.

In a gang of floor layers there ought to be one that understands the work *well*, so as to direct the rest and not allow any carelessness or inattention to the details of the work. As a general thing too little attention is paid to the *details*, and a poorly laid floor is the result, and when it is down it is a hard thing to make right.

WOOD CARPET

Wood carpet consists of hardwood strips, 1½ inches wide, glued on a strip of cloth. By its use hardwood floors can be had at a moderate price. It can be fastened or laid by driving two small wire brads in each strip near the edge and in rows about 9 inches apart.

DISHED FLOORS

Church floors are now frequently made in a circular dished shape if the pews are arranged in circles; or if the pews are arranged straight, the floor is an inclined plane, so that every seat in the church will command a clear view of the chancel. The pitch in either case should not exceed ½ inch to the foot, as a greater inclination is unpleasant to walk upon.

If a room is to be placed under the room with the dished floor, frame the floor for an incline, and then nail on furring strips on the joists, so as to get the proper dished effect. These furring strips should be 2 inches wide and well nailed to the joists. At the outer walls the depth of these furring strips will be considerable, and if they have to exceed 6 inches in depth, they should be bridged the same as joists, to prevent buckling. Use as large timber as possible, that is, if a depth of 6 inches is needed, do not build up out of 2″×2″ stuff, but use 2″×6″, etc.

The furring strips may be either nailed across the joists or nailed on top of them lengthwise.

In regard to arranging the girders and joists, we will quote from Kidder's *Building Construction and Superintendence :*

"If the girders supporting the incline run the same way as the inclination, they should be given the same pitch as the floor, and the joists will then be level from end to end. If the girders run in the opposite direction, they will be level endwise, and the joists will be on an incline. Whether the joists or girders shall be inclined depends upon the plan of the room, the openings in the walls, and the desired spacing of the columns. In arranging girders it should be remembered that it is better, and generally more economical, to give the longer span to the joist and to limit the girder spans to 12 or 13 feet for wood and 16 feet for steel.

"Joists 14 inches deep should not be less than 2½ inches thick, as 2-inch joists are apt to fail by buckling unless bridged every 4 or 5 feet by solid bridging, and such bridging will usually cost more than the extra thickness of the joists."

Another method is to set the joists radially and at the proper pitch, the floor being laid directly on the joists, no furring being required. The girders in this case are short and tangent to circles struck from the centre used for the circular pews, and one line higher than the other, so as to give the proper pitch to the joists when fitted and placed on top.

To fit these joists is a simple matter, as they are practically the same as rafters, and it is only necessary to find the foot cut to have them fit properly on the girders so as to give the required pitch to the floor.

On account of the number of short girders used in this method, the space beneath the room can hardly be used for school room, etc., owing to the columns.

The floor is nailed directly to the joists or furring strips and is laid in straight lines across the room, as if it were level, as the boards will spring enough so as to get the dished shape unless the inclination be too great, when a number of them will have to be shaped. Of course the ends of the boards will have to be cut at a mitre, so as to fit nicely against the wall. The boards should be well nailed so as to prevent the tendency to spring back to a level position.

VENEERING

The use of veneers is rapidly growing, and special appliances are now made for the economical handling of veneers; these appliances are invaluable in millwork factories, but they, however, are too costly for the carpenter who has but an occasional job to do. The following directions by Mr. John Phin give the necessary hints:

The softest woods should be chosen for veneering upon—such as common cedar or yellow pine. Perhaps the best of all for the purpose is "arrow board," 12-foot lengths of which can be had of perfectly straight grain, and without a knot. Of course no one veneers over a knot. Hard wood can be veneered—boxwood with ivory, for instance—but wood that will warp and twist such as nasty cross-grained mahogany, must be avoided.

The veneer, and the wood on which it is to be laid, must both be carefully prepared, the former by taking out all marks of the saw on both sides with a fine toothing plane, the latter with a coarser toothing plane. If the veneer happens to be broken in doing this, it may be repaired at once with a bit of stiff paper glued upon it on the upper side. The veneer should be cut rather larger than the surface to be covered; if much twisted, it may be damped and placed under a board and weight overnight. This saves much trouble; but veneers are so cheap—about 2 cents a foot—that it is not worth while taking much trouble about refractory pieces. The wood to be veneered must now be sized with thin glue; the ordinary glue-pot will supply this by dipping the brush first into the glue, then into the boiling water in the outer vessel. This size must be allowed to dry before the veneer is laid.

We will suppose now that the veneering process is about to commence. The glue in good condition, and boiling hot, the bench cleared, a basin of hot water with the veneering hammer and a sponge in it, a cloth or two, and everything in such position that one will not interfere with or be in the way of another.

First, damp with hot water that side of the veneer which is not to be glued; then glue the other side. Second, glue over as quickly as possible the wood itself, previously toothed and sized. Third, bring the veneer rapidly to it, pressing it down with the outspread hands, taking care that the edges of the veneer overlay a little all round. Fourth, grasp the veneering hammer close to the pene (shaking off the hot water from it) and the handle pointing away from you; wriggle it about, pressing down tightly, and squeezing the glue from the centre out at the edges. If it is a large piece of stuff which is to be veneered, the assistance of a hot flatiron from the kitchen will be wanted to make the glue liquid again after it has set; but do not let it dry the wood underneath it, or it will burn the wood and scorch the veneer, and ruin the work. Fifth, having got out all the glue possible, search the surface for blisters, which will at once be betrayed by the sound they give when tapped with the handle of the hammer; the hot iron (or the inner vessel of the glue-pot itself, which often answers the purpose) must be applied, and the process with the hammer repeated.

When the hammer is not in the hand, it should be in the hot water.

The whole may now be sponged over with hot water, and wiped as dry as can be. And observe throughout the above process never to have any slop and wet about the work that you can avoid. Whenever you use the sponge, squeeze it well first. Damp and heat is wanted, not wet and heat. It is a good thing to have the sponge in the left hand nearly all the time, ready to take up any moisture or squeezed-out glue from the front of the hammer.

So much for laying veneers with the hammer, which, though a valuable tool for the amateur, is not much used in the best cabinet-makers' shops. Cauls are adopted instead. They are made of wood the shape and size of the surface to be veneered; or, better still, of rolled zinc plate, and, being made very hot before a good blaze of shavings, they are clamped down on the work when the veneer is got into its place. They must be previously soaped to prevent them sticking to the veneer. The whole is then left to dry together.

The hammer is quite sufficient for most amateurs. I have laid veneers with it 5 feet long by 18 inches wide without assistance, and without leaving a blister. Cauls, however, are very necessary if a double-curved surface has to be veneered, or a concave surface; they need not be used for a simple convex surface. By wetting well one side of the veneer, it will curl up, and can easily be laid on such a surface; but it will be well to bind the whole round with some soft string to assist in keeping it down while drying.

A writer in *The Woodworker* says that the claim has been made that by using oak and other fine woods in the form of veneer, the figure or grain shows up better than when the wood is used in regular lumber form. This of itself is not a fair statement—it leaves too much room for misunderstandings. It is, in the first place, foolish to state that the figure will show up better in a thin piece of veneer than it will on an inch board of the same character. But, on the other hand, we can, in making a column, for example, by using veneer, get the fine figure all the way around, while if it were made from a solid piece, it would only show on two sides. In other words we do not get a better figure in veneer, but we can make a better display with it.

INLAYING

Every one has noticed that in ordinary inlaying there is a very ugly glue joint, equal in its width to that of the saw used, which runs round the whole of the inlaid pattern. This, of course, looks bad, and further, it involves the use of a very fine saw to reduce the width as much as possible. This, again, involves the use of comparatively thin wood. To avoid this, tilt up the saw-table a little on one side—say the right; with it in this position, cut out the right side of a letter—say a capital I; obviously the uppermost of the two pieces of wood on which we are operating would have its I slightly broader than the bottom one. Then finish the letter, being always careful to make the cut "sun about," as the phrase is, i.e., in the same direction as the hands of a clock move. We now have an I cut out of the top piece slightly broader and longer than that cut out of the lower one; if we have proportioned the amount of "tilt" of the table, with due regard to the thickness of the saw and of the wood used, the upper I will just fit neatly and tightly into the space left in the lower piece. Apply plenty of glue and gently tap the letter or monogram into its place, and we have a glue joint which will be barely visible. The amount of slope required in the table is very slight, and one soon finds out the happy medium.

ROOF FRAMING

The entire subject of roof framing is simple if one understands the underlying principles. Of course any one acquainted

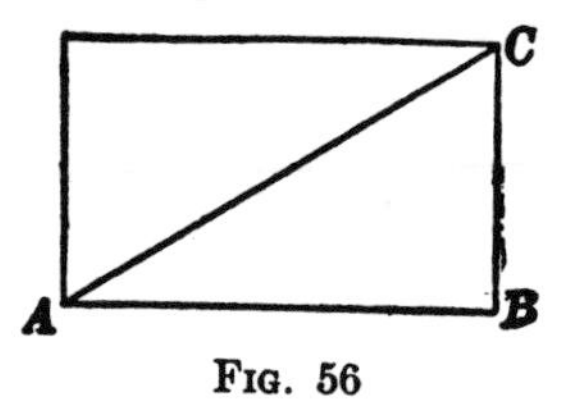

Fig. 56

with descriptive geometry can work out any problem, but the whole subject of framing, as far as the ordinary carpenter is concerned, can be understood by a knowledge of the relation of the diagonals of a rectangle and a cube to the sides.

The diagonal of a rectangle is a very simple thing, and any one who is able to calculate its length and find it by drawing has really a thorough understanding of how to get the various cuts and lengths for common rafters.

Fig. 56 shows the diagonal of a rectangle. It can be easily drawn. Its length can be figured by adding together the squares of the length and width of the rectangle and extracting the square root of this sum; that is to say, that ABC is a right angle, and the square described on the hypothenuse AC is equal to the sum of the squares described on the other sides, AB and BC.

From this figure we also find the angles made by the diagonal at A and C.

This principle can be applied to laying out common rafters, as shown in Fig. 57. The rectangle and its diagonal may be drawn on a piece of paper to scale, or on the floor. The run of the roof is the length of the rectangle, and the rise the width. The plumb cut and foot cut are had at once.

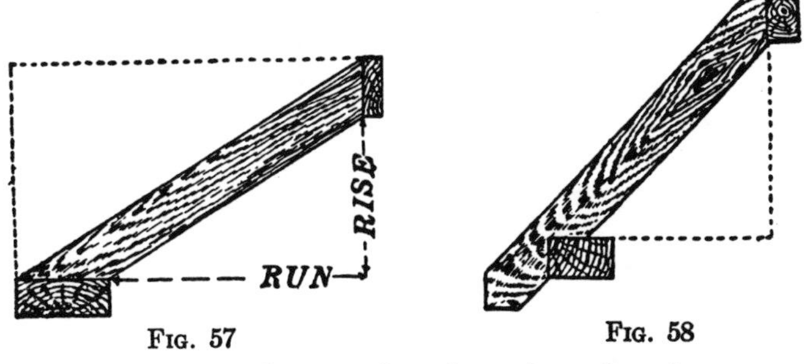

FIG. 57 FIG. 58

These angles may be transferred to the rafter by means of a bevel or by the geometrical method of describing one angle equal to another angle.

In Fig. 57 the rafter is shown with its top on the diagonal of the rectangle; when the rafter is notched, as is often the case, the diagonal of the rectangle is taken as the middle of the rafter, half of the rafter being placed on each side of the diagonal, as shown in Fig. 58, which clearly shows the angles for both cuts.

Fig. 59 shows a cube. From the upper front corner to the

lower back corner of opposite sides is drawn the diagonal of the cube. To find the exact length of this, the following method may be used:

Draw a square, ABCD, Fig. 60, equal to one side of the cube; then draw its diagonal, CB. On this diagonal construct a rectangle, CBEF, the sides CE and BF each being equal to a side of the cube; the diagonal (EB or CF) of this rectangle is the diagonal of the cube.

Remember that the diagonal CB is the base of the diagonal of the cube or the run of the hip.

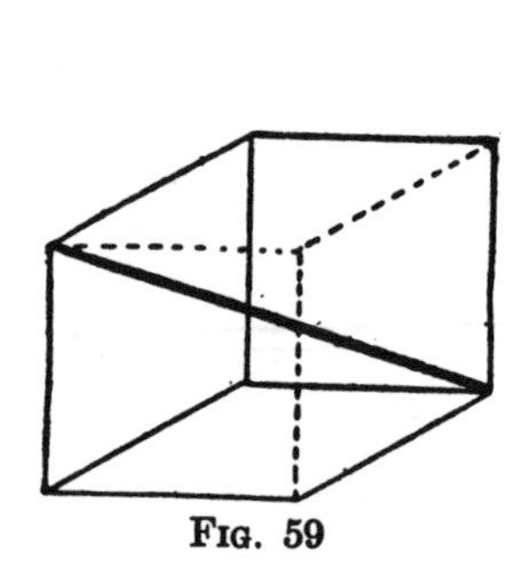

Fig. 59

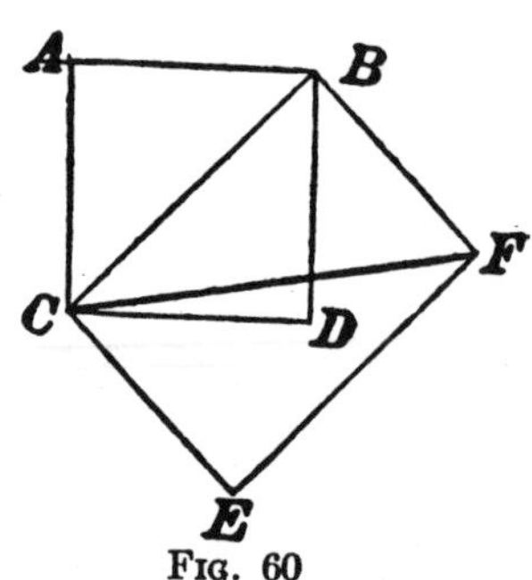

Fig. 60

The principle given above can be applied to find the length of hip or valley rafters, also to find the cuts of these rafters.

Suppose it was necessary to find the length of the rafters in the roof shown in Fig. 61. The rise of the roof is 5 feet; the length of the common rafters is found as described above. To find the length of the hip rafter, notice that the distance between the foot of common rafter and the corner on which the hip rafter rests is 5 feet (one-half of 10 ft.), and as the rise is 5 feet, the hip rafter is the diagonal of a 5-foot cube.

Draw Fig. 60 to a scale so that AC, CE, etc., represent 5 feet; then measuring CF by the same scale will give the length of the hip rafter. The bevel for the foot cut is obtained at F, for the plumb cut at C.

The same principle can be applied to valley rafters, also hoppers.

A line perpendicular to CE, Fig. 60, and ending at the diagonal CF will be the length of a jack or valley rafter at that point.

The angle CFE gives the side cuts of the jacks against the hips. This angle will also give the cuts for valleys, as a valley is but a hip turned upside down.

The plumb cut for the jacks is the same as for the hip, that is, the angle ECF.

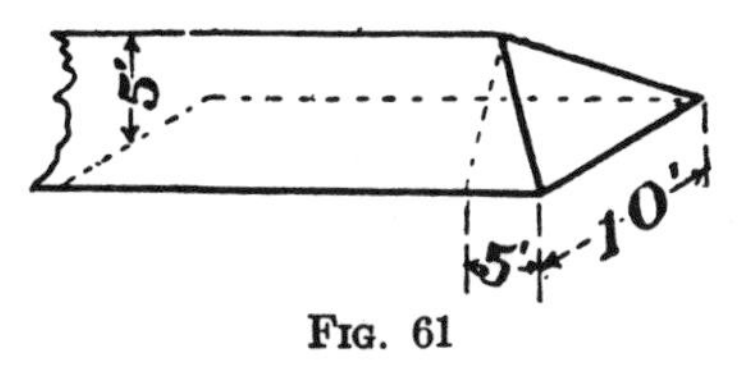

FIG. 61

All of the above can be easily proven by drawing the diagrams on a piece of paper and cutting them out and folding them together at the hip line. It will be found that the correct form of the hip will be formed.

So far the principle has been explained, but this is rather a long way to follow in practical work. The best short-cut method is by the use of the steel square.

Any one understanding about diagonals, as just explained, will have no trouble in using the square accurately and with confidence, as he will know *why* as well as *how* to use it.

Just place the square on any of the diagrams and you will have the numbers on tongue and blade which will give the various cuts.

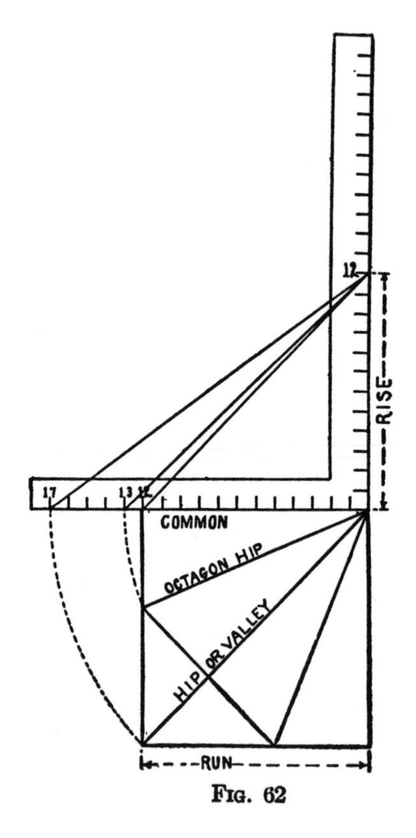

FIG. 62

For the sake of simplicity, the run is always taken as one foot, and the rise a certain number of inches, depending on the pitch.

For half pitch it would be 12 on tongue and 12 on blade.

For third pitch it would be 12 on tongue and 8 on blade.

For quarter pitch it would be 12 on tongue and 6 on blade, etc.

The mark on blade gives the plumb cut and the mark on tongue gives the foot cut.

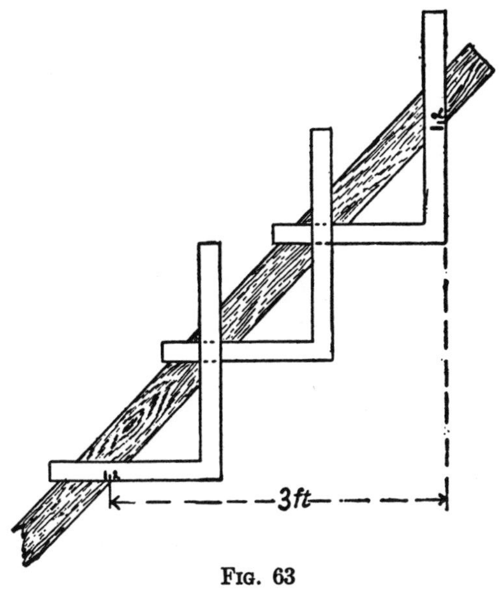

FIG. 63

Fig. 62 illustrates the idea. For common rafters take 12 on tongue and the rise on blade.

For hips or valley take 17 on tongue and the rise on blade (17 because that is the base of a hip rafter where the common rafters have a run of 12 inches).

For octagon hips take 13 on tongue and rise on blade.

For side cut of hip take 17 on tongue and length of hip on blade. Cut on blade.

For side cut of jacks take 12 on tongue and length of common rafter on blade. Cut on blade.

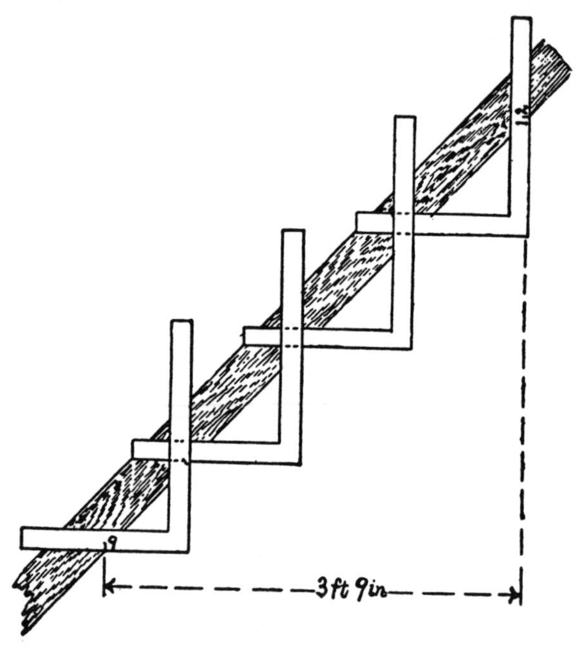

FIG. 64

For side cut of octagon jacks take 5 on tongue and length of jack on blade. Cut on blade.

For backing the hip rafter take rise on tongue and length on blade. Cut on tongue.

The reason 12 inches is used for the run is because the length of rafter can be easily laid off without any figuring, the square being applied as many times as there are feet in the run, as shown in Fig. 63.

Where the run is given in feet and inches, the method is the same as shown in Fig. 63, only that in the last application of the square the run is taken as the odd number of inches, instead of 12. Thus Fig. 64 shows the application in laying out a rafter with a run of 3 feet 9 inches.

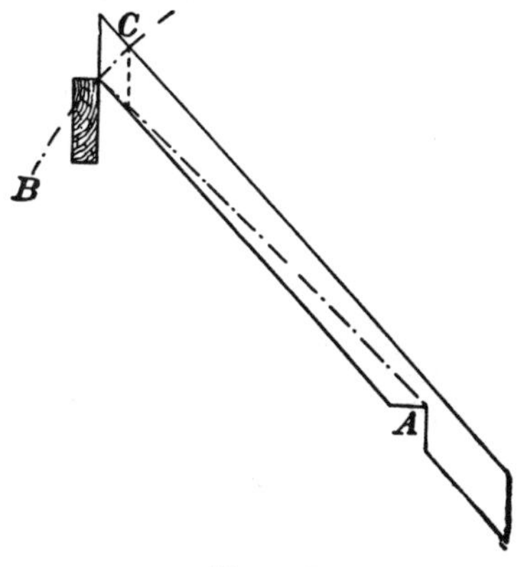

Fig. 65

When a rafter is elevated and found to be too long to go in position, the simple method devised by Mr. F. A. Williams, and shown in Fig. 65, can be used to get the right length without any calculating.

Take AB for a radius and strike an arc cutting the rafter at C. Mark the plumb cut at this point, and it will be the required length.

Another way is to square across the rafter at the top of the ridge and make a new plumb cut which is near enough to be practical.

BRACES

The same rules relating to common rafters are used to get the length of braces; simply lay the steel square on the required number of times to get the required run, and you will have length and cuts; see Fig. 63.

HOPPERS

A hopper is practically a hip roof turned upside down, and the same rules for getting cuts and lengths of hips are used. The following directions are by Mr. Arnold Huisenfeldt:

Take Figs. 66 and 67 as mitre cuts when the sides are flat

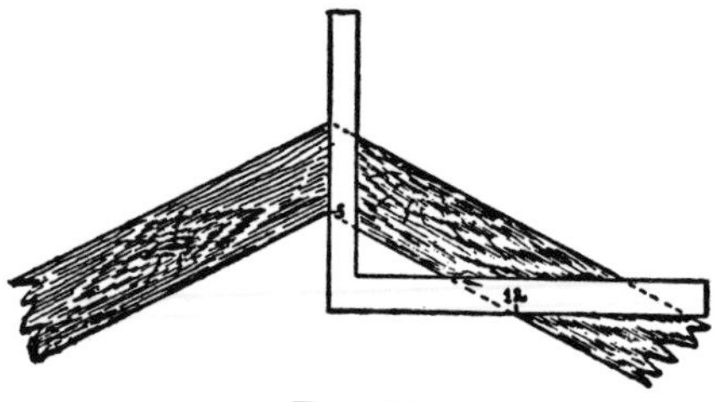

FIG. 66

or level, that is, when they have no rise. Fig. 66 is for an octagon mitre, Fig. 67 for a square one.

Now for hoppers or anything similar, to find the face bevel for the flat side of boards, provided the rise all around is the same, take the hypothenuse of the rise per foot by the mitre bevel per foot.

Thus, if we wish to make a square hopper with a 9-inch rise per foot, the mitre, as shown by Fig. 67, is 12×12 when the sides are flat. The hypothenuse of 9 inches (the rise per foot) and 12 inches (the run) is 15; now taking 12 and 15 on the square, 12 gives cut for face of sides.

To find the bevel for the edges for mitre cuts, take the reversed pitch, which in the above case rises 16 inches per foot; the hypothenuse of 16 and 12 is 20; therefore the mitre bevel is 12 and 20; give cut mark on 12 side.

For octagon hoppers, 5 and 12 give the bevel when sides are flat (see Fig. 66). Now to find bevel when the rise is 9 inches per foot; the hypothenuse of 9 and 12 is 15; therefore 5 and 15 give the cut for face of sides.

For bevel of edges of mitre cut, take the reversed pitch per foot run. For the above the reversed pitch rises 16 inches per foot; the hypothenuse of 16 and 12 is 20; therefore 5 and 20 give the cut: mark on 5 side.

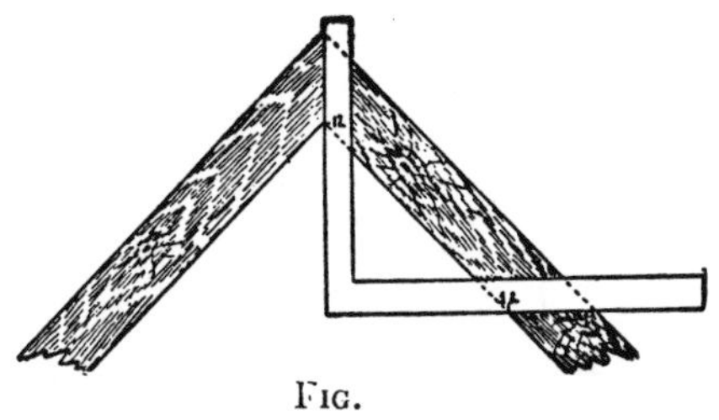

FIG.

The above method will work on any polygon, provided the rise is the same all around. The mitre marks for various polygons can be found in various books on the steel square.

DESIGN FOR WHEAT BIN

A form of wheat bin which may be made large or small, built of any size lumber, and which will never leak, is shown in Fig. 68, taken from the *American Miller*.

Build the hopper first. Put in the rafters, then floor them, running the flooring crosswise and having it extend out past where the studding will be. Cut the studding on a bevel to fit the hopper. The sketch is an end view.

SAWING OFF FENCE POSTS ON A BEVEL

Take two pieces of ⅞″×6″×3′, and nail together like a corner board. Then saw the required bevel on one end and use it for the top. Then nail the template to the post, just plumb two ways, and you will find it an easy matter to saw the posts on the required bevel. Then change the template to the next

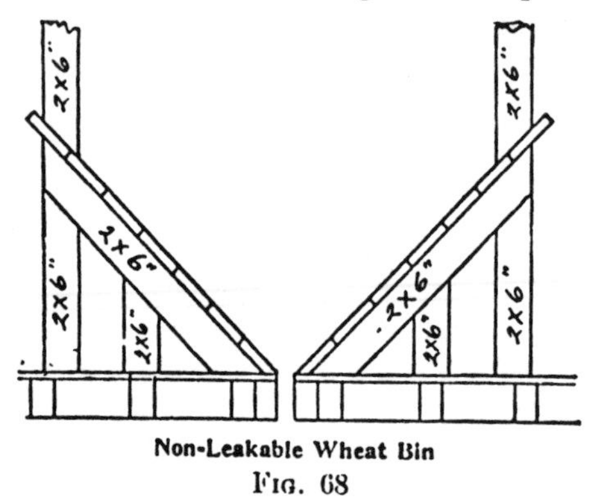

Non-Leakable Wheat Bin

FIG. 68

post, and so on. This makes a guide for the saw, and the posts will come very accurate, says F. A. Williams.

A PRACTICAL SPAR OR FLAG-POLE RULE

First shape the pole the desired taper. For example, if the stick is 6″×6″, apply the octagon rule.

Place the 24-inch steel square across the 6″×6″, as shown in Fig. 69; mark on 7, which will be 1¾ inches from outside.

Next make a template out of ⅛-inch pine, 5″×12″, as shown

in Fig. 70, and insert a lead pencil 1¾ inches from inside cut. Then place template across the 6″×6″, and slide it to the tapering end, holding it all the way, so the pole fills the inner space. Plane off the corners and round the pole to finish. Submitted by F. A. Williams.

WOOD CURB FOR WELLS

Perhaps my method of putting together a round well curb may interest some of your readers, says Dayton B. Halderman.

Here in the West we use wooden curbs mostly, and I build a round curb in the following manner:

In the first place, curbs should not be made over 8 feet long, because they would be hard to handle.

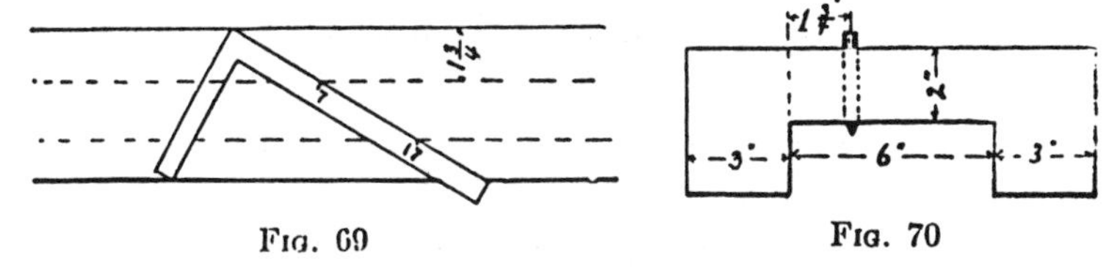

FIG. 69 FIG. 70

I cut my boards the length I wish the curb to be, and level them as if they were for a barrel or tank,—a little full on the inside.

Then lay the boards on a pair of long trestles, edge to edge, and with the outside or feather edge up; now tack a cleat across each end as a temporary stay; next, draw a wire across the boards (this will form a hoop) 12 inches from each end and one in the middle, leaving the wires project over on each side about 12 inches. The wires should be stretched tightly across the boards and stapled about an inch from both edes of each board. Fig. 71 shows it complete.

Now "up end" it in a perpendicular position by edge of well, having timbers laid across the well, and remove the cleats. Then swing the outside around over the well until the ends meet, and twist the wires together; toe-nail the boards a little where they need it. The curb is now standing over the well, ready to lower.

Two men can lower it very easily by driving two pegs in the ground 2 feet apart on one side of the well. Around these pegs

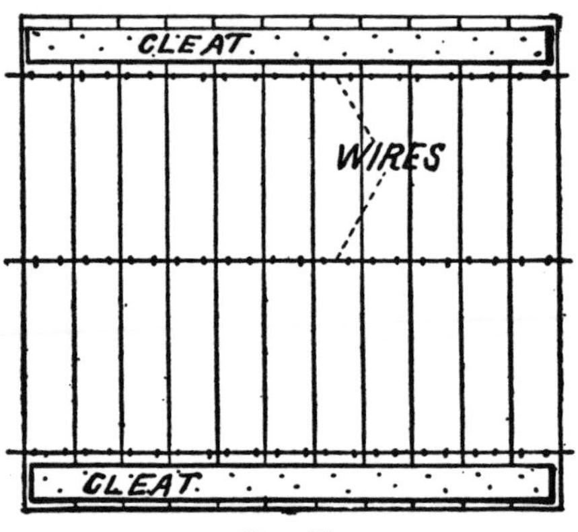

FIG. 71

a rope is passed, the rope also going under the curb to a man on the opposite side, who holds the curb up while the other man removes the timbers from over the well and keeps the curb steady. The curb is gradually allowed to slide down in the well; when about half way down, the rope is loosened and it drops to the bottom.

If the well is deep, several curbs can be set one upon another from the bottom to the top.

The illustration will show the method of construction quite clearly.

A well 3 feet in diameter will take eleven boards 10 inches wide. The bevel may be found according to the scheme shown in Fig. 72.

THE BEVEL OF A TANK STAVE

One not versed in the contrivance of round tanks is apt to waste a lot of time in getting the bevel for the staves, says a writer in *The Woodworker.* Indeed, cases might be cited of men of ripe experience in that line, who find it necessary to make a

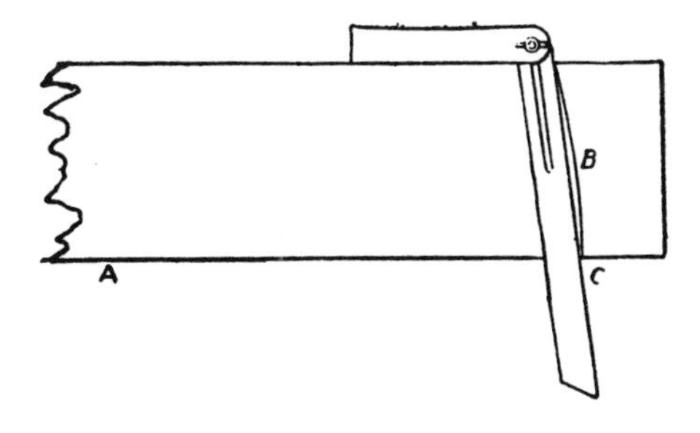

FIG. 72

rather elaborate diagram for the purpose. All this work is needless, for one has only to take one of the pieces intended for a stave and, with the trammel set to the radius used in striking the bottom, or if one wishes to be minutely correct, to that of the outside of the tank, and from a point at one edge of the stave, as at A in Fig. 72, strike the curve B. The bevel set to touch the two ends of this curve will be correct, but it is usually the practice to set it a little back at C, so that the staves will be a trifle open on the outside, which opening closes when the tank is wet and the staves spring to the curve of the bottom.

INDEX

A

B

C

D

F

G

H

R

S

V

W

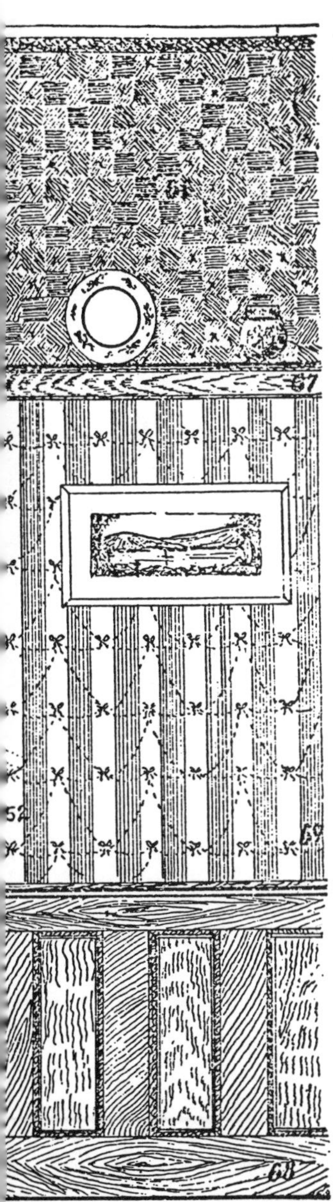

To get this drawing within a reasonable size, so that it can be presented in this book, it was necessary to group the various designs together, so as to show the different styles of trim, etc. Some of the details are exaggerated in size compared with the scale of the drawing, so as to show them plainly. The names of the various parts are:

1. Window-sill
2. Sub-sill
3. Furring
4. Quarter rounds
5. Sill-cap
6. Header
7. Pocket or opening to weight box
8. Blind-hanging stile or exterior casing
9. Exterior sash stop
10. Clap-boarding, shingles, or outside covering
11. Sheathing or roof boarding

11^{1}. Sheathing paper

12. Stud
13. Laths
14. Plastering
15. Architrave, interior casing, or window trim
16. Stop bead
17. Parting bead or strip
18. Pulley stile
19. Sash weights
20. Window latch or sash lock
21. Pulley
22. Sash cord
23. Meeting rail of outside sash
24. Meeting rail of inside sash

24^{1}. Bottom rail of sash

24^{2}. Top rail of sash

25. Stop bead
26. Window head
27. Exterior casing
28. Sash stile
29. Astragal or sash bar
30. Sash lift
31. Window pane or glass
32. Panel back or breast
33. Base block
34. Window trim, casing, or architrave
35. Corner block
36. Window trim

36^{1}. Cap trim for window

37. Picture molding
38. Dado
39. Wall
40. Border
41. Stile
42. Hanging stile
43. Top rail
44. Middle rail
45. Lock rail
46. Bottom rail
47. Muntin
48. Panels

48^{1}. Upper panel

48^{2}. Middle panel

48^{3}. Lower panel

49. Knob
50. Keyhole
51. Base blocks
52. Door trim
53. Head block
54. Door trim
55. Cap trim for door
56. Carving
57. Hinges
58. Ornamental casing
59. Door saddle or threshold
60. Header
61. Door stop or jamb mold
62. Door jamb
63. Furring
64. Door jamb
65. Plaster cornice
66. Frieze
67. Plate rail
68. Wainscoting
69. Wall covered with paper
70. Skirting or base board
71. Chair rail

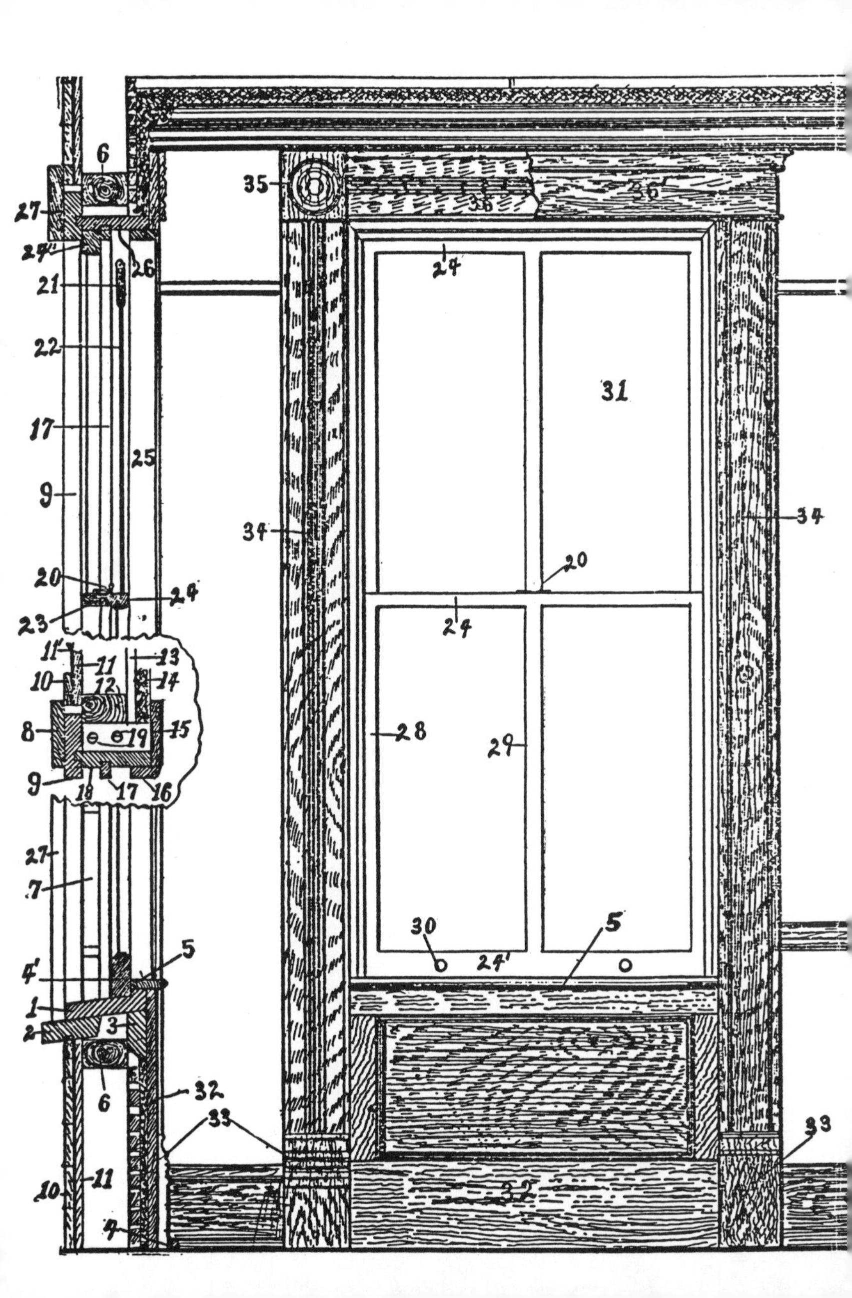

6
35
36
36
27
24'
26
24
21
22
31
17
25
9
34
34
20
20
24
23
24
11'
11
13
10
12
14
8
19
15
28
29
9
18
17
16
27
7
30
24'
5
5
4'
1
2
3
6
32
33
33
10
11
32
4

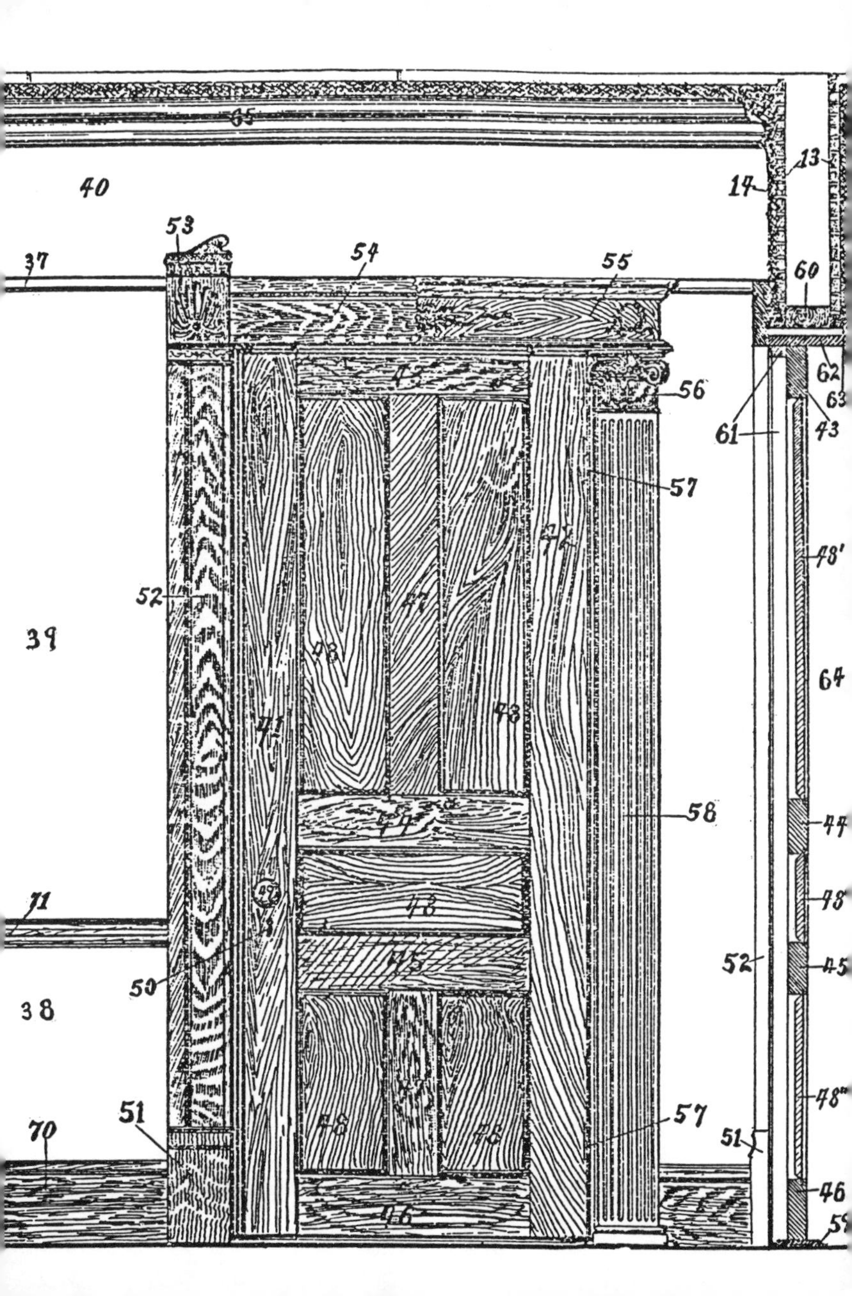

65
40
14
13
53
37
54
55
60
47
56
62
63
61
43
57
52
39
41
48
47
48
42
48'
64
49
58
44
71
48
48'
50
45
52
45
38
57
48''
51
70
48
46
51
46
46